To the Memory of Joel W. Chastain, Jr.

CONTENTS

PREFACE

This book relates topics on which I have spent a significant portion of my technical life. This book blends my academic training with many of my research activities of the past twenty-five years. I took most of my undergraduate work in engineering physics (at the Ohio State University), and my graduate work in physics (also at the Ohio State University), and spent much of my career in nuclear engineering, a field in which I am currently teaching at the University of Cincinnati. From 1955 to 1960 I was Operating Supervisor of the Battelle Research Reactor, during which time span reactor research emphasized radiation effects on materials and components. I have lectured on much of the material in both undergraduate and graduate classes in the Nuclear Engineering Program at the University of Cincinnati.

Joel W. Chastain, Jr., to whose memory this book is dedicated, had a large influence on my early professional life at Battelle Memorial Institute and also on my personal life. He played a significant role in the development of a number of topics. I deeply regret that his life ended before this material was in print.

I extend my appreciation to my former colleagues at Battelle Memorial Institute who participated in much of the research herein reported, and to my colleagues at the University of Cincinnati for their patience and understanding. A number of graduate students at the University of Cincinnati have assisted in preparing sections of this text. Foremost is Mr. Lun Cheng, my teaching assistant in the radiation effects courses I have taught both at the University and at the Zimmer Nuclear Power Station during the past several years. He collected and organized my lecture notes to form the backbone of this text. Mr. Ahmed A. Moosa contributed significantly to Chapter 7 (Radiation Effects on Steel) as did Mr. Cecil Staud to Chapter 11 (Biological Effects of Radiation). One of my colleagues, Dr. Alvin Shapiro, contributed information on gamma dosimetry (Chapter 12). The research work of a number of my graduate students is also reflected, and referenced, throughout the text. My deep appreciation is extended to my wife, Jan, who not only worked patiently in seeing this text through to completion, but spent many hours in preparing all the (over 100) drawings for this text. A special note of thanks to Dolores Kappes, Charlene Oaks and Karen Newman for their excellent performance and diligence in preparing the manuscript. Finally, my thanks is given to the staff and associates of Research Dynamics Incorporated, Cincinnati, Ohio, for their encouragement in the publishing of this book.

J. N. Anno

1 *INTRODUCTION*

With the advent of the "atomic age," which has been given inaugural dates from December 2, 1942, the first sustained fission chain reaction at the CP-1 reactor, to July 16, 1945, the explosion of the first A-bomb, has come a more urgent need for engineering knowledge on the effects of radiation on materials. With the transition after World War II from warfare to civilian use of nuclear reactors, much needed to be learned about materials for use in constructing nuclear power plants. Test reactors sprang up across the world, the most famous and one of the earliest being the Materials Testing Reactor (MTR) at Arco, Idaho. The major question facing designers of reactor structural materials was "how much damage will the radiation do to the materials?" Consequently, early writings on the subject of this text became known under the category of radiation damage. This stigma still pervades the nuclear engineering world.

However, a wider scope of the subject is more properly categorized as radiation effects, which includes the vast listing of the beneficial effects of radiation. In fact, the beneficial effects preceded the "damage" effects by many years. The medical benefits of radiation in cancer treatment are well-known. Every modern hospital today has not only access to radiation therapy equipment, but administers radioisotopes routinely for diagnostic purposes. Industry quickly recognized the beneficial uses of radiation in a wide variety of applications from thickness gauges to radiography of welds.

Consequently, this book is entitled, "Notes on Radiation Effects" rather than "Notes on Radiation Damage." Although nuclear medicine and industrial uses of radioisotopes are not treated per se, some writings are included which could be categorized as beneficial effects of radiation. In a broader scope, all effects considered for the development of nuclear power plants can be categorized as beneficial!

This text is not intended to be a comprehensive treatment of all aspects of radiation effects. Rather, it is a collection of notes previously used in a presentation of the field on a survey basis. A number of very specialized writings have preceded this text, and the reader is encouraged to pursue the many references for detailed knowledge.

However, the uniqueness of this text is that it presents an overall coverage of the more useful engineering topics encountered in a study of radiation effects.

The entire field of nuclear radiation is a fascinating one, the sub-topic of radiation effects being spawned by man's curiosity and desire to put to use the discoveries of his predecessors. In an editorial sense, and to place this sub-topic in perspective, a brief review of the history of the nuclear world reveals the long path from "earth, fire, and water" to "rads and rems."

1.1 The Nuclear World -- An Historical View

One wonders what events really heralded or ushered in the nuclear world. The authors prefer to believe that the approach to the nuclear world began as early as the fifth century B.C. The "atomic school" of Greek philosophers: Democritus, Leucippus and Lucretius, believed that if one kept subdividing a piece of material, ultimately one would reach a smallest unit which could no longer be subdivided -- the atom. This school of thought survived through many centuries, through the Dark Ages, with little being added. At the end of the Dark Ages the alchemists, and later the chemists, began classifying matter into basic substances. By the time of the American Civil War, the Russian chemist D. I. Mendeleev had classified all known matter in the form of elements in the well-known periodic table. The idea was accepted that elements consisted of basic building blocks called atoms -- the smallest unit of matter. Or was it?

Towards the end of the nineteenth century, scientists, notably Sir J. J. Thompson, began chipping away at the atom. In 1897 he showed the existence of electrons as fragments of an atom, and further, he showed that these fragments were the same no matter what atom from which they came. In this sense, the atom was first split in 1897.

In 1911 Sir Ernest Rutherford showed by bombarding matter with small charged particles (alpha particles, which are emitted by radium -- Curie's discovery of 1898 -- among others) that the atom had a very small core or <u>nucleus</u> which was positively-charged. He proposed that the electrons whirled around this nucleus in orbits. By 1913 Neils Bohr completed the modern "classical" model of the atom: a tiny central nucleus, positively-charged, around which electrons whirled in discrete orbits. Most of the atom was found to be made up of empty space. The almost weightless electrons whirl around an incredibly dense but tiny nucleus. From Rutherford's experiments, and knowing atomic weights, one calculates the density of the nucleus to be 1.2 x 10^{14} gm/cm^3 -- 10 trillion times as dense as lead!

It remained for Chadwick in 1932 to complete the picture of the nucleus with the discovery of the neutron as a basic particle. The neutron was discovered by bombarding beryllium with alpha particles:

$$^{9}_{4}Be + ^{4}_{2}He \longrightarrow ^{12}_{6}C + ^{1}_{0}n \qquad (1.1)$$

It was then formulated that the nucleus consisted of two types of basic particles: positively-charged protons and neutral neutrons. But what glued them together?

We must now retreat a bit to the year 1905 in which Albert Einstein formulated the famous law that every schoolboy now recites:

$$E = mc^2 \qquad (1.2)$$

If mass disappears or is annihilated, energy must appear according to this formula. Conversely, if energy disappears, mass must appear. In other words, mass and energy have an equivalence. They are just different forms of something -- to coin a word, "massergy."

By various means, scientists have determined the masses of the basic particles: protons, neutrons and electrons.

proton mass	=	1.007277 amu	
neutron mass	=	1.008665 amu	
electron mass	=	0.000549 amu	(1/1836 proton mass)
sum		2.016491 amu	

where 1 amu = 1 atomic mass unit = 1/12 mass of ${}^{12}_{6}C$ atom = 1.6604×10^{-24} gm. Now the simplest atom containing all three basic particles is the deuterium atom (heavy hydrogen), consisting of a neutron bound to a proton as the nucleus, with a single electron in orbit. The experimentally-derived mass of the deuterium atom is 2.01410 amu. Yet when we add the masses above of the individual constituents we get 0.00239 amu more than the measured value of the deuterium atom where all are put together. This extra mass, Δm, termed the "mass defect," has been converted into "binding energy" to glue the nucleus together according to

$$\text{Binding Energy} = \Delta mc^2 \quad (1.3)$$

The larger the nucleus (the more the number of protons and neutrons) the more binding energy it has. One can therefore visualize that a very heavy nucleus, such as uranium, has a tremendous store of binding energy.

In 1938, as a result of studies of bombarding uranium with neutrons, isotopes of barium and krypton were found. Barium is so far away from uranium in terms of mass that Hahn and Strassman in Germany suggested that some uranium atoms had actually split. The reaction might be, for example,

$${}^{238}_{92}U + {}^{1}_{0}n \rightarrow {}^{239}_{92}U \rightarrow {}^{?}_{56}Ba + {}^{?}_{36}Kr + \text{neutrons} \quad (1.4)$$

This possible fragmentation of a nucleus was termed "nuclear fission" by Meitner and Frisch, colleagues of Neils Bohr. Just by guessing at mass numbers for the isotopes of barium and krypton, it was calculated that enormous amounts of energy would be released in the fission process. The fact that more neutrons are given off in a fission process than are required to initiate it suggested to Enrico Fermi in 1939 that a chain reaction might be possible.

Then came World War II! The military implications of a chain reaction with tremendous energy release sparked the United States to establish the Manhattan Project -- a project that was to cost $2 billion and result in the atomic bomb.

But first it had to be established that a chain reaction was possible. Fermi and his associates set about to build the first nuclear reactor -- the Chicago Pile-1 -- under the stadium at the University of Chicago. On December 2, 1942, the first manmade self-sustaining nuclear chain reaction was achieved, the reactor being operated at a power of 0.5 watts. Thus was born a new era for mankind -- to bring both destruction and construction.

The Manhattan Project continued secretly at a fervid pace. A date remembered more than December 2, 1942, is July 16, 1945, at 5:30 a.m. when the first atomic explosion was achieved. Frequently, this is called the start of the Atomic Age. The event was witnessed by a renowned group of scientists and military men gathered in the desert approximately 120 miles southeast of Albuquerque, New Mexico. The rest of the atomic bomb story is generally well-known. On August 6, 1945, President Truman announced that an atomic bomb had been dropped on Hiroshima with "an effect hitherto unapproached in war."

Fortunately peace -- at least a temporary peace -- followed the dramatic illustration of nuclear forces at

Hiroshima and Nagasaki. Scientists were then free to pursue the peaceful uses of nuclear energy. Many experimental and research reactors were built across the United States. After much materials testing, fuel development, and design work, power reactors for the peaceful production of electrical power came into being. The first large-scale electrical power-producing reactor was the EBWR, the Experimental Boiling Water Reactor, designed, built and operated by the Argonne National Laboratory in 1956 with an output of 5,000 kWe. Today modern plants have output of the order of 1,000,000 kWe. It is projected that by the year 2000 A.D., less than twenty years from now, a considerable fraction of the electrical power in the United States will come from nuclear power plants.

What else is on the horizon? Problems with the operation of fission reactor power plants, the accumulation of obnoxious radioactive fission products being among them, have been widely publicized. Yet we desperately need a long-term source of energy to conserve our dwindling fossil fuels. The author considers the fission reactor as a thankful bridge between power from fossil fuel and the nuclear power of the future -- fusion power. Consider the fusion reactions with the isotopes of hydrogen: deuterium (D) and tritium (T).

$$
{}^2_1D + {}^2_1D \begin{cases} \nearrow {}^3_2He + {}^1_0n + 3.27\ \text{MeV} \\ \searrow {}^3_1T + {}^1_1H + 4.03\ \text{MeV} \end{cases} \qquad (1.5)
$$

$$
{}^2_1D + {}^3_1T \longrightarrow {}^4_2He + {}^1_0n + 17.6\ \text{MeV}
$$

These reactions are "clean" and deuterium is abundant in the ocean waters. In fact, there is sufficient deuterium in the oceans (about 10^{14} tons) to supply the world with energy for a billion years at the projected 2000 A.D. rate of consumption. The problem facing engineers and scientists is to harness this energy in a fusion reactor.

And in this sense, the Nuclear World is yet an infant.

1.2 The Contents

The material presented in this text should be useful to both fission and fusion reactor engineers, as well as those engineers engaged in other specialized radiation effects studies. The following three chapters essentially lay groundwork for the remaining topics of the book. Two of these chapters, Chapters 3 and 4, present basic material on the interaction of nuclear radiations with materials. Much of this material can be found in other texts on nuclear engineering. However, the foundation is repeated here, augmented with a slant toward the topic of this text.

Chapter 5 presents the first engineering use of the prior material: the transient effects of radiation on electrical components. The full impact of this topic is being discovered in currently active research on simulation of fission power plant accident conditions. Chapter 6 presents useful engineering design data and thinking on the effects of radiation on organic materials. Chapter 7 takes the same approach on the effects on steel. However, since the reader may not have an intimate knowledge of materials science and terminology, Chapter 7 also presents introductory material on stress-strain relations and fatigue life phenomena.

Chapter 8 is devoted entirely to the fission process and to selected radiation effects in fission reactor fuels. Emphasis is placed upon the thermal spike model and fuel swelling. Chapter 9 covers an often-neglected area of radiation effects, namely surface effects. Much of this material has been only sparsely published and should be of interest to those readers with predominantly research interests.

Chapter 10 is primarily speculative, and ventures a few of the author's thoughts on potentially directly beneficial effects of nuclear radiation. The speculations, however, are supported by experimental observations.

Although the primary thrust of this text is the effects of radiation on materials and components, for completeness Chapter 11 briefly presents the effects of radiation on biological organisms. Man has forever been exposed to a radiation environment and must understand how to safely harness this tool for his use. Concluding this text is a chapter on measurement of neutron and gamma radiation. The first half of this final chapter is devoted to dosimetry for experiments on radiation effects, whereas the concluding remarks consider dosimetry for personnel working in artificial radiation environments.

Where appropriate, throughout the various chapters numerical illustrations are included to assist in the use of the material presented in the text. For academic use, a list of problems is included at the end of each chapter.

2 UNITS AND GENERAL SCALE OF RADIATION DAMAGE

In order to make meaningful calculations in radiation effects problems, it is necessary to have a set of units in which radiation fields and radiation effects can be expressed quantitatively.

2.1 Roentgen

The oldest and, earlier, most accepted unit was the Roentgen, R, which uses the ionization in air as its basis. The definition of this radiation unit is as follows: "1 Roentgen = that quantity of X- or gamma ray radiation such that the associated corpuscular emission per 0.001293 g of air produces, in air, ions carrying 1 electrostatic unit (1 esu) of electricity of either sign." The amount of energy deposited in air by a Roentgen of X- or gamma rays can be calculated in the following manner. The unit electrical charge, i.e., the electronic charge, is 4.80×10^{-10} esu, and this is the quantity of electricity carried by each member (positive or negative) of an ion-pair. Consequently, $1/(4.80 \times 10^{-10})$, i.e., 2.08×10^{9}, ion-pairs are required to give a total charge of 1 esu of either sign. Hence, the absorption of 1 R of gamma radiation in 0.001293 g of air results in the formation of 2.08×10^{9} ion-pairs in air. The energy required to produce one ion-pair in air is known to be about 34 eV and, so, the energy required for the formation of 2.08×10^9 ion-pairs is $34 \times 2.08 \times 10^9 = 7.07 \times 10^4$ MeV or 0.113 ergs. The energy absorbed per gram of air per Roentgen is thus 0.113/0.001293 = 87.7 ergs.

It must be emphasized that the concept of using the Roentgen as an exposure unit applies only to X-rays or gamma rays in air at a point outside the body. Consequently, the Roentgen unit is not useful in a technical sense both to describe radiation interactions taking place within a body, and for other types of radiation.

2.2 Rad

A radiation effect is also a function of how much energy is deposited at the point of interest. An absorbed dose unit, the rad, is defined as the quantity of radiation leading to the absorption of 100 ergs of energy per gram of irradiated material. (In soft tissue, an exposure of 1 R leads to the absorption of approximately 1 rad which also can be used as a crude approximation for engineering problems.) This more fundamental absorbed dose unit is more useful in discussing radiation effects, since it is independent of both the type of radiation and the material being irradiated.

2.3 Gray (Gy)

The modern SI unit for absorbed dose is the Gray (Gy) unit which is the absorption of 1 joule/kg of the irradiated material.

One Gy unit of dose thus is equivalent to $\frac{10^7 \text{ ergs}}{1000 \text{ gm}} = 10^4$ ergs/gm = 100 rads.

While the Gray unit is the designated SI unit for dose,

the rad unit is still the most commonly used unit in both the literature and in the field. The conversion, stated above, is a simple matter. This text will generally use the more widely accepted rad unit of dose.

2.4 Dose Threshold for Various Materials

The functional radiation dose threshold for various materials is given in Figure 2.1 [Reference (2-1)].* This simple chart is surprisingly useful to the engineer for "rule of thumb" estimates of radiation effects.

Practically every property is affected by radiation. We are interested in which properties are most affected and by how much they are affected. Some indication of the effects of neutron radiation on various materials is given in Figure 2.2 [Reference (2-2)]. The integrated fast neutron flux, or neutron fluence, is simply the product of neutron flux and time. The units of fluence are neutrons per cm^2, or simply, (nvt).** A further indication is given in Figure 2.3, which gives the level of neutron exposure which appreciably alters some representative materials and renders them useless [Reference (2-2)].

Note that the very approximate conversion factor between epithermal fluence (nvt) and rad dose is

$$1 \text{ rad dose} \cong 10^9 \text{ epithermal neutron fluence.}$$

---

* References at end of chapter.

** The concept of neutron flux is discussed in the following chapter.

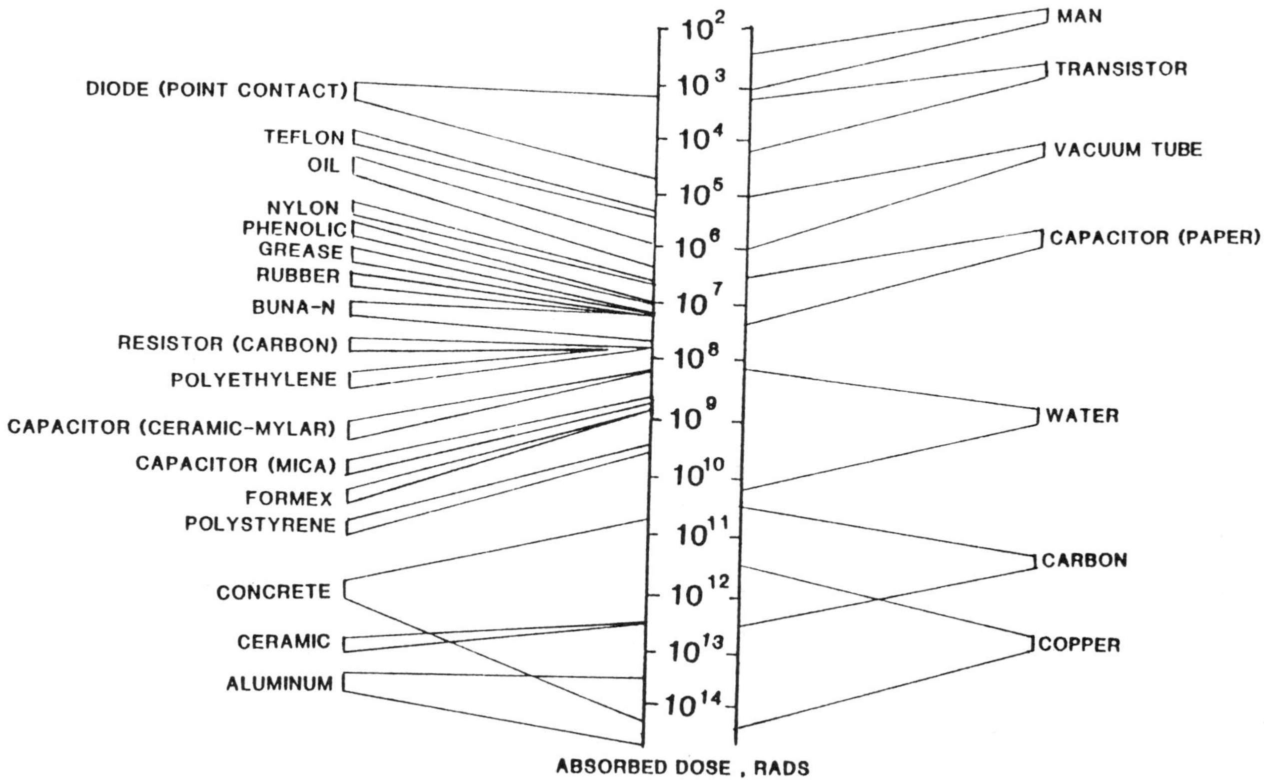

FIGURE 2.1 Functional radiation dose thresholds.

[Reproduced with permission from Radiation Effects Information Center (REIC), Memorandum 21, "Space Radiation and Its Effects on Materials," Available from the National Technical Information Service, Springfield, VA 22161, as AD261277]

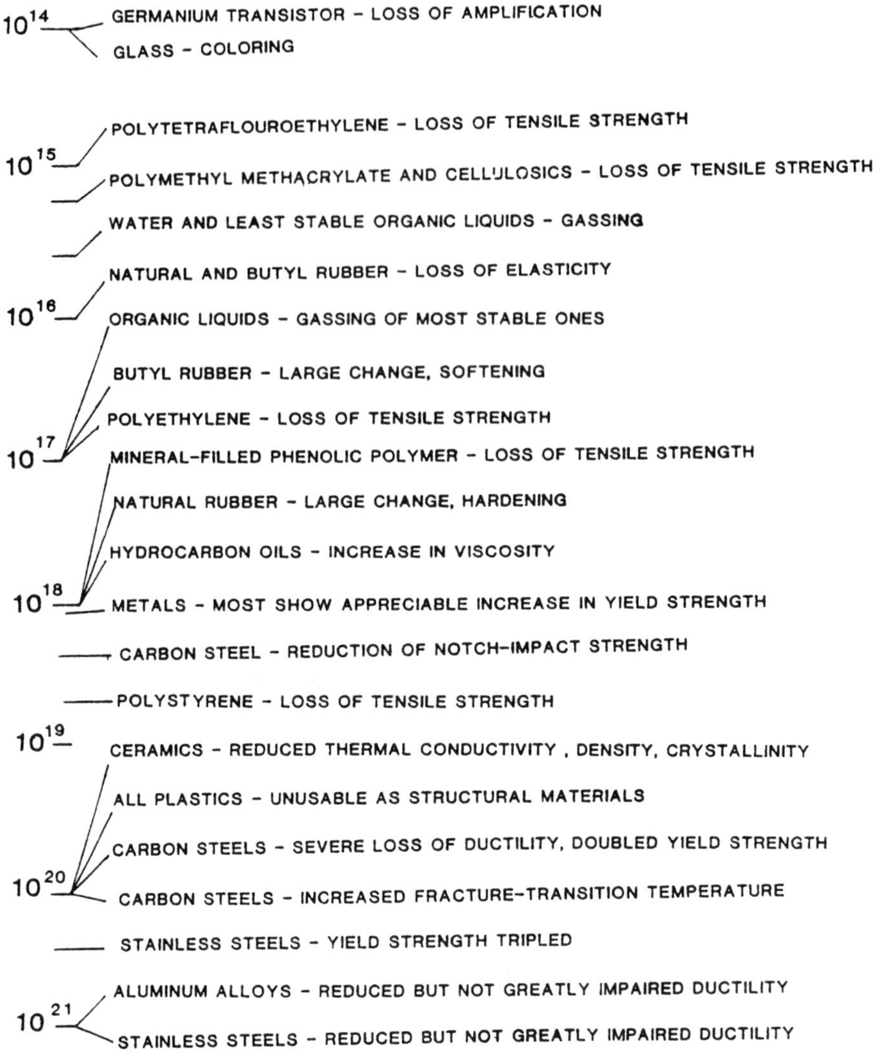

FIGURE 2.2 Effects of neutron radiation on various materials. Indicated exposure levels are approximate. Indicated changes are at least 10%.

[Reproduced with permission from Charles O. Smith, Nuclear Reactor Materials, Addison-Wesley Publishing Company, Inc., Reading, Mass., 1967, Ref. (2-2)]

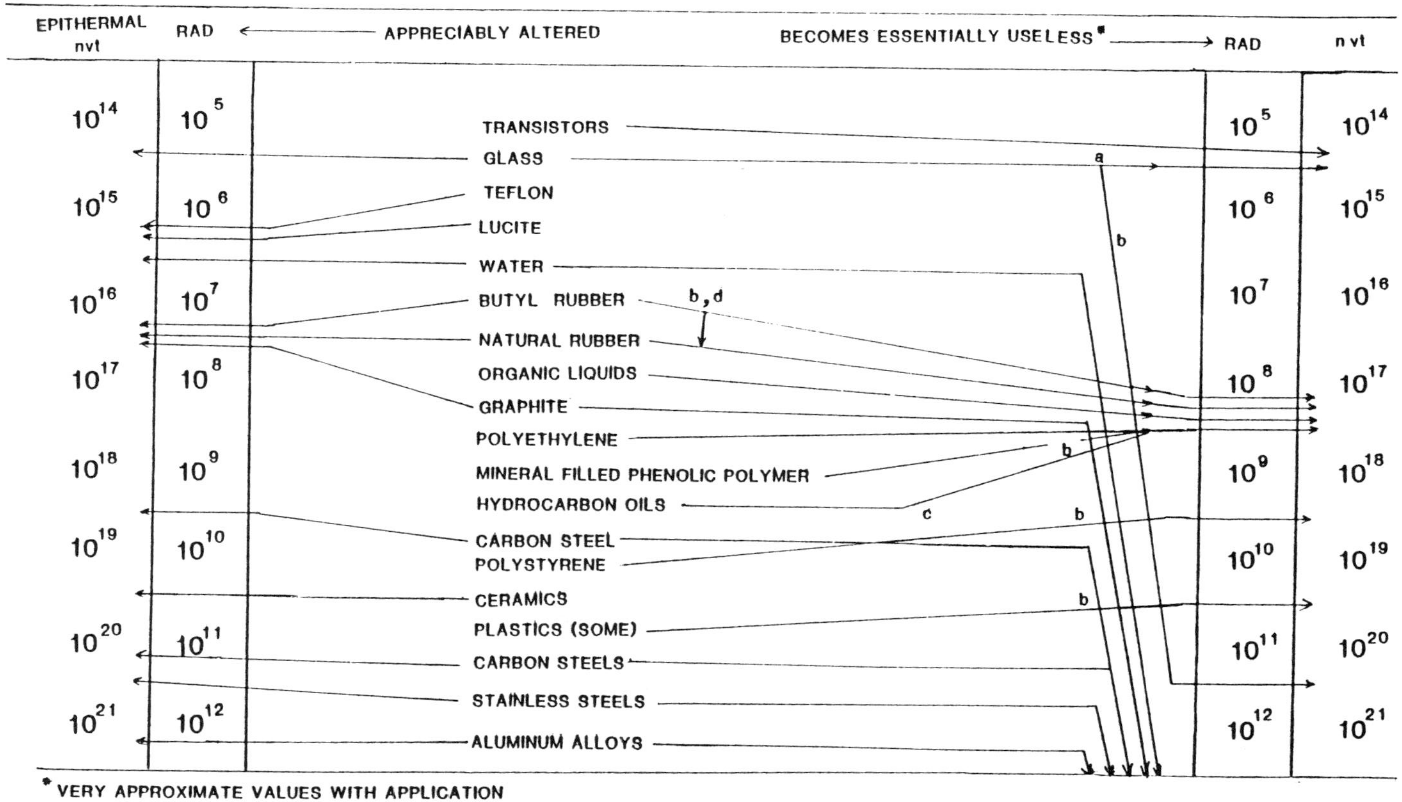

FIGURE 2.3 Radiation stability of various substances: (a) transparent medium, (b) as lubricant, (c) structural material, (d) electrical insulators. [From Smith, Ref. (2-2)]

REFERENCES

(2-1) REIC Memorandum 21, "Space Radiation and Its Effects on Materials," Radiation Effects Information Center, Battelle Memorial Institute, Columbus, Ohio, June 30, 1961, Available from the National Technical Information Service, Springfield, VA 22161, as AD261277.

(2-2) Smith, Charles O., Nuclear Reactor Materials, Reading, MA, Addison-Wesley Publishing Co., Inc., 1967.

PROBLEMS FOR CHAPTER 2

1. Carbon steel is exposed to the average fast neutron flux of 10^{14}n/(cm^2)(sec). How long will it take to double the yield strength of this piece of carbon steel?

2. An O-ring made of Buna-N is contained in some machinery located near the reactor pressure vessel and is exposed to a "modest" gamma dose rate of 100 rads/hr. Approximately, how long will the O-ring "last" until it suffers radiation damage?

3. An engineer tells you he wants to use concrete in a radiation environment such that it will receive 10^{16} rads. What is your advice on the possibility of radiation damage to the concrete?

4. A piece of carbon steel is exposed to a fast neutron flux of 10^{13}n/(cm^2)(sec). How many days until it will suffer severe loss of ductility?

5. A material receives an epithermal neutron irradiation (fluence) of 10^{19}nvt. Approximately what rad dose has it received?

3 *FUNDAMENTAL INTERACTIONS OF NEUTRONS WITH MATERIALS*

There are various types of nuclear radiations (radiations originating in the nucleus). However, for practical purposes the most important are neutrons and gamma rays. This and the next chapter of the text present discussions of the fundamental interactions of these two types of radiation with materials from an engineering viewpoint. For a more basic physics discussion the reader is referred to Reference (3-1).

3.1 Neutron Interaction Theory

Of basic interest to the nuclear engineers is the reaction rate of neutrons with materials. It is the reaction rate that determines the power level in a nuclear reactor, or the damage to a material exposed to radiation. Usually to determine the reaction rate, one needs to know the probability of an interaction in the presence of a neutron, usually expressed through the "neutron cross section" of a material.

3.1.1 The neutron. The neutron is, to the best of our knowledge, ascribable as a neutral particle with mass of 1.008665 amu (0.13% larger than the proton). A free neutron is radioactive with a half-life of 12.8 ± 2.5

minutes. The free neutron splits up into a proton, an electron and a neutrino. However, when bound in the nucleus the neutron is stable and considered a basic particle.

The existence of the neutron was suggested by Rutherford in 1920 but not demonstrated until 1932 when Chadwick showed its existence as a result of research on the disintegration or transmutation of nuclei by alpha particles.

3.1.2 Neutron reactions. Because the neutron has no electric charge it is a very penetrating particle and readily undergoes reaction with the nuclei of atoms, i.e., the neutron is not subject to the repulsive electrostatic forces in the neighborhood of a positively-charged nucleus. The reaction between a neutron and a nucleus gives rise in most cases to an alpha particle, a proton, a gamma ray photon, or to two neutrons. The atom remaining after the reaction is a different nuclide either of the same element or another element, depending on the reaction. In many cases this new nuclide is radioactive and may decay with the emission of a gamma ray, a beta particle, or both. (A beta particle is either an electron ${}^{0}_{-1}\beta$ or a positron ${}^{0}_{+1}\beta$). By detecting the radioactivity of the material and with a knowledge of the half-life or decay constant of the material, the reaction rate of the neutrons with the material can be deduced.

As an example, a neutron reaction with gold, a common material used to measure "flux," is the (n,γ) reaction:

$${}^{1}_{0}n + {}^{197}_{79}Au \longrightarrow {}^{198}_{79}Au^{*}$$

It is to be noted that both charge and mass numbers are conserved in nuclear reactions. The compound nucleus

$^{198}_{79}Au^*$ is an excited state and almost instantaneously disintegrates:

compound nucleus $\rightarrow$ residual nucleus + ejected particle

$^{198}_{79}Au^* \longrightarrow {}^{198}_{79}Au$ + 9.5 MeV (in the form of gamma rays; maximum energy of a gamma ray in this reaction is 6.494 MeV)

The residual nucleus, in this case $^{198}_{79}Au$, undergoes radioactive decay which 99% of the time (i.e., "branching ratio," B.R., is 0.99) is of the form

$$^{198}_{79}Au \xrightarrow[\text{half-life}]{2.7 \text{ days}} {}^{0}_{-1}\beta + {}^{198}_{80}Hg + 0.411 \text{ MeV gamma ray}$$

By observing the number of gamma rays of this energy emitted from the radioactive gold one can determine the amount of ^{198}Au which is present, and hence the reaction rate of the neutrons with ^{197}Au. (The beta particles could also be counted to give the same results.) There are many other materials which undergo reactions useful for determining neutron reaction rate. Cobalt is good for "high flux" environments.

3.1.3 The sources of neutrons. In nuclear fission reactors, the source of most neutrons is that of fissioning of ^{235}U (or ^{239}Pu). About 84% of all neutron captures by ^{235}U result in fissioning of the uranium, with the production of 2.5 new neutrons on the average. In the other 16% of captures by ^{235}U, stable ^{236}U is formed. The fission neutrons are fast neutrons (high velocity) with an average energy of about 2 MeV. The actual energy spectrum of the fission neutrons is represented within experimental error by Watt's empirical formula

$$n(E)\, dE = 0.484\, e^{-E} \sinh \sqrt{2E}\, dE \qquad (3.1)$$

where n(E) dE = number of neutrons between the energies E and E + dE, normalized to unity, and E is in the unit of MeV. This formula is good from about 0.075 to 17 MeV, although with little error the upper limit to the fission neutrons energy is generally taken as about 10 MeV (see Figure 3.1). As can be seen from this graph, the distribution peaks below 2 MeV, i.e., the most probable energy is less than the average energy, being 0.72 MeV.

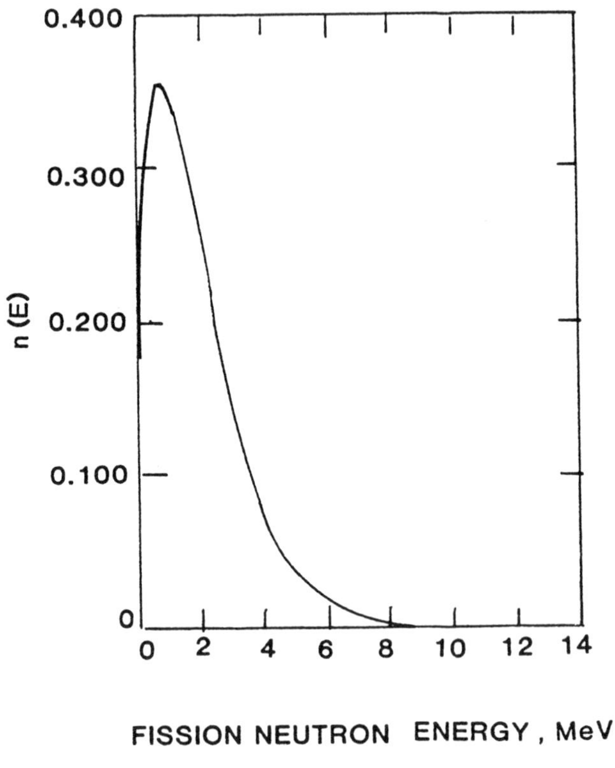

FIGURE 3.1 Fission neutron energy spectrum.

The fast neutrons born at fission collide with the reactor materials, in particular the "moderator," and through loss of energy at each collision are slowed down to very slow or "thermal" neutrons which bounce around in thermal equilibrium with the environment. In the process of slowing down many neutrons escape from the core (leakage neutrons) and others are absorbed by the core materials. Those neutrons which escape leakage or absorption have, in the process of slowing down, an energy distribution given approximately as

$$n(E)\, dE \propto \frac{dE}{E^{3/2}} \qquad (3.2)$$

Neutrons which have been slowed down to thermal energies continue to diffuse until absorbed by the reactor materials, or until leaked out the sides of the reactor. The thermal neutrons may be considered analogous to a gas in thermal equilibrium with the environment, with a Maxwellian velocity distribution.

$$n(v) = 4\pi n_0 \left(\frac{m}{2\pi kT}\right)^{3/2} v^2 e^{-\frac{mv^2}{2kT}} \qquad (3.3)$$

where n_0 = total thermal neutron density, m = mass of the neutron, k = Boltzmann's constant, and T = absolute temperature. With this distribution

v_0 = most probable neutron velocity = 2198 m/sec at 293°K (energy = 0.0253 eV)

and

$\bar{v}$ = average neutron velocity = $\frac{2}{\sqrt{\pi}} v_0 = 1.128\, v_0$.

In terms of energy, the Maxwellian distribution [Equation (3.3)] can be written as

$$\frac{dn}{n} = \frac{2}{\sqrt{\pi}} \frac{E^{1/2}}{(kT)^{3/2}} e^{-E/kT} dE \qquad (3.4)$$

where dn is the fraction of particles having energy between E and E + dE. Defining the dimensionless parameter

$$u = \frac{E}{kT} \tag{3.5}$$

the Maxwellian distribution can be written in more compact form as

$$\frac{dn}{n} = \frac{2}{\sqrt{\pi}} u^{1/2} e^{-u} \, du \tag{3.6}$$

This becomes the weighting factor for averaging quantities over a thermal neutron distribution.

Illustration 3.1 Find the average value of the energy of a thermal neutron. Define $\overline{u} = \frac{\overline{E}}{kT}$, then

$$\overline{u} = \frac{\int_0^\infty u \frac{dn}{n}}{\int_0^\infty \frac{dn}{n}}$$

But $\int_0^\infty \frac{dn}{n} = 1$, i.e. the distribution of Equation (3.6) is normalized. Thus, $\overline{u} = \int_0^\infty u \frac{dn}{n} = \int_0^\infty \frac{2}{\sqrt{\pi}} u^{3/2} e^{-u} du = 3/2$, so that $\overline{E} = 3/2\ kT$.

These thermal neutrons do very little damage to materials, per se, because of their extremely small energy. However, they can cause nuclear reactions, the products of which may be damaging.

3.1.4 Reaction rate. Of principal interest in radiation effects is the reaction rate of neutrons with material. Consider a medium in which neutrons are present in a steady state motion. Looking at a cubic centimeter (unit volume) of this medium, suppose there is a number n(v) neutrons at each velocity v in this volume. [The neutron density at v is n(v) neutrons/cm^3.] The neutrons may be traveling in any direction. Suppose, for the sake of illustration, as sketched in Figure 3.2, that we have one neutron traveling at velocity v_1, three neutrons traveling at velocity v_2, and two neutrons traveling at velocity v_3. In time t, the total path traveled by this collection of neutrons is given by (path is the product of velocity and time)

$$\begin{array}{l}\text{total path traveled by}\\ \text{collection of neutrons}\\ \text{in unit volume in time } t\end{array} = (1)\,v_1t + (3)\,v_2t + (2)\,v_3t \tag{3.7}$$

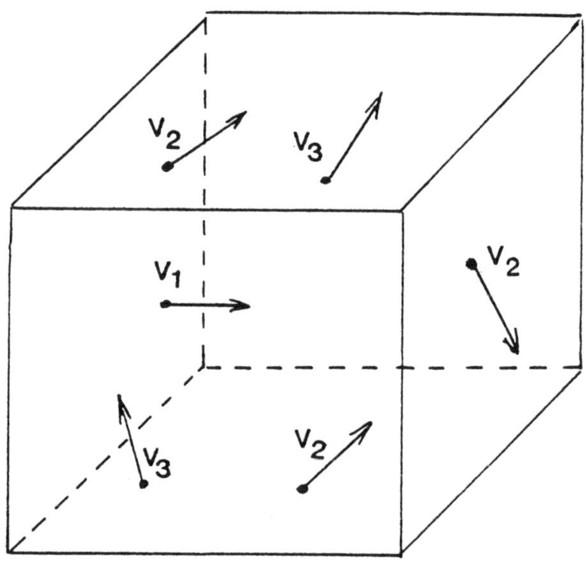

FIGURE 3.2 Unit volume of randomly-moving particles.

This relation can be generalized for a larger collection of neutrons by summing over the number with velocity v_i, i.e.,

$$\text{total path traveled by generalized collection of neutrons in unit volume in time } t = \sum_i n_i v_i t \qquad (3.8)$$

Thus, the total path traveled by the neutrons in this unit volume per unit time is obtained by dividing Equation (3.8) by the time t,

$$\text{total path traveled by neutrons per unit volume per unit time} = \sum_i n_i v_i \qquad (3.9)$$

Now consider the fact that the total number of neutrons n_T in this unit volume is

$$n_T = \sum_i n_i \qquad (3.10)$$

and divide and multiply Equation (3.9) by n_T

$$\begin{aligned}\text{total path traveled by neutrons per unit volume per unit time} &= \frac{n_T}{n_T} \sum_i n_i v_i \qquad (3.11)\\ &= n_T \left(\frac{\sum_i n_i v_i}{\sum_i n_i} \right) = n_T \bar{v}\end{aligned}$$

where $\bar{v}$ is the average velocity of the entire collection of neutrons, defined by

$$\bar{v} = \frac{\sum_i n_i v_i}{\sum_i n_i}$$

or in the integral form

$$\bar{v} = \frac{\int\limits_{\text{all } v} n(v)\,v\,dv}{\int\limits_{\text{all } v} n(v)\,dv} \tag{3.12}$$

Now, if λ is the average distance a neutron travels between interactions (sometimes called the "mean free path"), then the interaction rate R is simply

$$R = \begin{array}{l}\text{number of interactions}\\ \text{per unit volume}\\ \text{per unit time}\end{array} = \frac{\begin{array}{l}\text{total path traveled by}\\ \text{neutrons per unit}\\ \text{volume per unit time}\end{array}}{\begin{array}{l}\text{average distance a neutron}\\ \text{travels between interactions}\end{array}}$$

$$= \frac{n_T \bar{v}}{\lambda} \tag{3.13}$$

The determination of the mean free path between interactions can be illustrated by considering a beam of one type of particles incident on a "slab" of cross-sectional area A containing another type of particle as illustrated in Figure 3.3 (a). Let I_0 be the intensity of the incident beam on the slab and let n_s be the number of particles per unit volume of the slab material (i.e., the density of the particles in the slab). As the beam interacts with the particles in the slab, collisions occur and some of the beam particles are removed from the direct beam. At a distance x into the slab, the initial intensity I_0 will have been reduced to intensity I, which depends upon how far the penetration has been, i.e., upon x. Now consider a very small thickness of the slab Δx located at x. Incident upon this small slab is intensity I, and exiting this small slab will be intensity $(I-\Delta I)$, that is, because of interactions within Δx, the intensity will be reduced by a small amount of ΔI. Now let us view the thin slab Δx normally as the beam would see it, as sketched in Figure 3.3(b). The beam, at distance x, sees a slab of "inverse Swiss cheese," with a group of particles, each with effective area of interaction of σ (called the "cross section"). The total area blocked out by these particles in Δx is $n_s \sigma A \Delta x$. (There are n_s particles per unit volume, and the volume of the small slab is $A\Delta x$, so that the total number of particles in the slab is $n_s A \Delta x$.) The probability for an interaction is simply the geometric ratio of the effective area blocked out by the particles to the total area A of the slab, i.e.,

$$\text{probability for an interaction in a slab of thickness } \Delta x \text{ and area } A = \frac{\text{area blocked out by particles in } \Delta x}{\text{total area of slab}} \tag{3.14}$$

$$= \frac{n_s \sigma A \Delta X}{A} = n_s \sigma \Delta x$$

Thus the change in intensity of the beam (which will be negative) is given as the product of how many are present at x (which is the intensity I) and the probability of interaction, i.e.,

$$\Delta I = -I\, n_s \sigma \Delta x \tag{3.15}$$

The calculus solution to this equation for the variation of the beam intensity with distance into the slab is

$$I(x) = I_0\, e^{-n_s \sigma x} \tag{3.16}$$

i.e., the attenuation is exponential. The average distance a particle travels before interaction (precisely the mean free path) is given by the calculus definition as

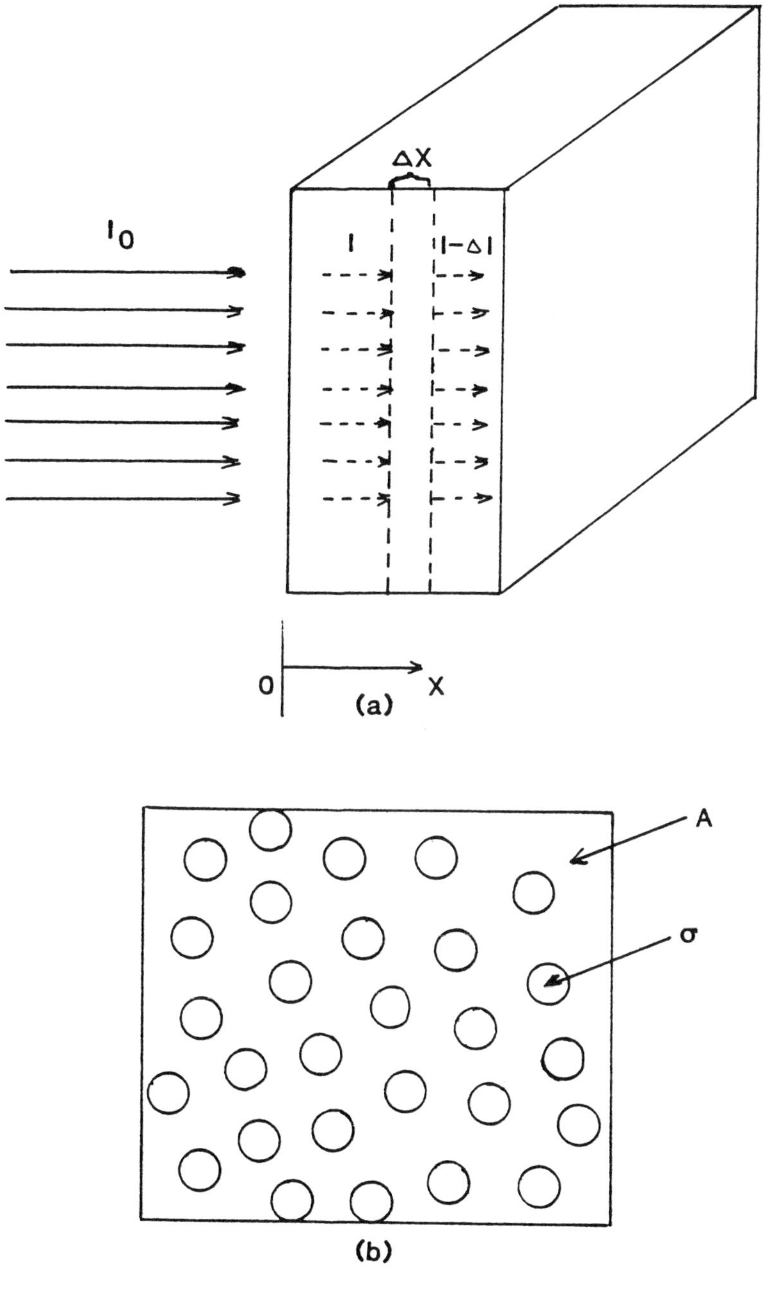

FIGURE 3.3 Geometry for attenuation of a beam of particles.

$$\lambda \equiv \bar{x} = \frac{\int_{\text{all space}} x I(x)dx}{\int_{\text{all space}} I(x)dx} = \frac{\int_0^\infty x I_0 \exp(-n_s \sigma x)dx}{\int_0^\infty I_0 \exp(-n_s \sigma x)dx} \tag{3.17}$$

$$= \frac{1}{n_s \sigma}$$

Although Equation (3.17) has been derived for the attenuation of a beam of particles, for the purpose of illustration, it is quite general for random collisions. It is here convenient to introduce the "macroscopic cross section,"

$$\Sigma \equiv \frac{1}{\lambda} = n_s \sigma \tag{3.18}$$

Thus, from Equation (3.13) the final form of the particle interaction rate per unit volume is

$$R = \frac{n_T \bar{v}}{\lambda} = n_T \Sigma \bar{v} = n_T \Sigma \frac{\int_{\text{all } v} n(v) v dv}{\int_{\text{all } v} n(v) dv} \tag{3.19}$$

The quantity [n(v)v] appearing in Equation (3.19) is defined as the neutron flux density, or commonly abbreviated as just neutron flux,

$$\Phi = \int_{\text{all } v} n(v) v dv \tag{3.20}$$

Since $n_T = \int_{\text{all } v} n(v)dv$, the reaction rate in Equation (3.19) can be written simply as

$$R = \Sigma\Phi \tag{3.21}$$

[It should be observed that the mathematically correct use of this simple formula for reaction rate requires that the macroscopic cross section be averaged according to

$$\bar{\Sigma} = \frac{\int_{\text{all } v} \Sigma(v) n(v) v dv}{\int_{\text{all } v} n(v) v dv} \;] \tag{3.22}$$

It should be pointed out that the discussion has been simplified greatly by not considering in detail the type of cross section being measured. For example, in the slab arrangement shown previously, with a very thin slab the cross section we would obtain is a total cross section which includes both the absorption and scattering cross section, $\sigma_T = \sigma_a + \sigma_s$ since each interaction, whether absorption or scattering from the parallel path, prevents a neutron from reaching the detector. By special experiments σ_a and σ_s can be separated. Assume, then, that by experiment one has complete knowledge of $\sigma(v)$, hence $\Sigma(v)$ and hence $\lambda(v)$. Graphs of $\sigma(v)$ for the elements may be found in such compilations as given in Reference (3-2). For simplification of calculations, cross sections are usually tabulated simply as thermal (slow) or epithermal (fast) neutron cross sections. Appendix A gives a brief tabulation of the cross sections in this manner.

Illustration 3.2 Calculate the mean free path of high energy neutrons in zirconium.

The mean free path of a neutron is

$$\lambda = \frac{1}{\Sigma} = \frac{1}{N\sigma}$$

where N = atomic density

$$= \frac{\rho N_A}{A_w} = \frac{6.44 \times 6.02 \times 10^{23}}{91.22}$$

$$= 4.25 \times 10^{22} \text{ atoms/cm}^3$$

σ = microscopic cross section of epithermal neutrons

$$= 6.2 \text{ barns} = 6.2 \times 10^{-24} \frac{\text{cm}^2}{\text{atom}}$$

(Appendix A).

Introducing these values gives

$$\lambda = \frac{1}{4.25 \times 10^{22} \times 6.2 \times 10^{-24}} = 3.8 \text{ cm}$$

Note that cross sections are usually tabulated in barn units, where 1 barn $\equiv 10^{-24}$ cm^2.

Illustration 3.3 Knowing that $\Phi = n_T \bar{v}$, calculate the total neutron density for 2-MeV (fission neutrons) at a fast flux of 2.5×10^{14} n/(cm^2)(sec). (This flux is typical of peak fast fluxes in a thermal power reactor.)

The velocity of a 2-MeV neutron is

$$\bar{v} = \sqrt{\frac{2E}{m}} = \left(\frac{2 \times 2 \text{ MeV} \times 1.6 \times 10^{-6} \text{ ergs/MeV}}{1.67 \times 10^{-24} \text{ g}}\right)^{1/2}$$

$$= 1.98 \times 10^9 \text{ cm/sec}$$

Thus $n_T = \Phi / \bar{v} = 2.5 \times 10^{14} / 1.98 \times 10^9 = 1.26 \times 10^5$ neutrons/cm^3.

3.1.5 Neutron cross sections. Typical of the behavior of the absorption cross sections for (n,γ) reactions is the behavior of the cross section for gold-197, sketched in Figure 3.4 [Reference (3-3)]. At low neutron energies, the cross section decreases with increasing velocity as $\frac{1}{v}$ (or $\frac{1}{\sqrt{E}}$). This low energy behavior is known as "$\frac{1}{v}$" behavior, and is a consequence of the deBroglie wavelength λ' of the neutron depending on $\frac{1}{v}$, since

$$\lambda' = \frac{h}{mv} \tag{3.23}$$

Thus the smaller the neutron velocity the larger the wavelength so that the neutron may be thought of as extending out a greater distance to interact with a nucleus.

This description is somewhat naive and a much more refined interpretation of the "$\frac{1}{v}$" dependency is given in References (3-4) and (3-5). At intermediate neutron energies, several resonances occur in the cross section superimposed on the $\frac{1}{v}$ behavior and the cross section becomes quite large over small energy intervals. The cross section is large at the resonance energy because the neutron kinetic energy at resonance is just that amount which will form a compound nucleus with an excitation energy equal to one of its virtual energy states (energy states beyond the dis-association energy of the compound nucleus). At higher energies the cross section begins to decrease as $\frac{1}{E}$. The reason for this behavior is shown by the Breit-Wigner model of the cross section and is also discussed in References (3-4) and (3-5).

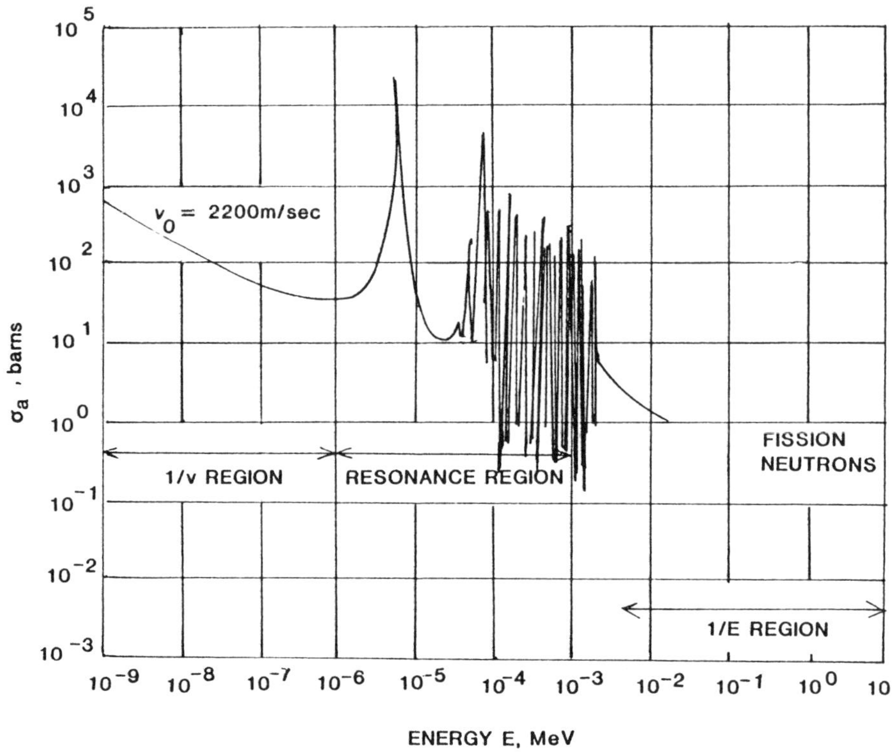

FIGURE 3.4 Sketch of (n,γ) cross section for gold-197.

From the gold (n,γ) cross section it is seen that the cross section is much higher at low neutron energies (in the thermal range). As a consequence of this behavior, most absorption reactions in "thermal" reactors (thermal fission >> fast fission) are due to the thermal neutrons. In some cases resonance reaction is significant. For most practical applications, in the thermal neutron range σ_a has $\frac{1}{v}$ behavior so that

$$\sigma_a = \frac{\sigma_0 v_0}{v} \tag{3.24}$$

where σ_0 and v_0 are any set of known velocity and known cross section at the neutron velocity. Incidentally, this $\frac{1}{v}$ (or $\frac{1}{\sqrt{E}}$) behavior implies a temperature dependency of the cross section of $\sqrt{\frac{293}{T}}$, where T is the Kelvin Temperature, when σ_0 and v_0 are evaluated at 293° K.

By replacing $\lambda(v)$ in the basic reaction rate Equation (3.13) the specific reaction rate can be written as

$$R(v) = \left(\frac{\rho N_A}{A_W}\right) n(v)\ \sigma(v) v \tag{3.25}$$

which is still exact. Now for thermal neutron absorption reactions, by substituting the $\frac{1}{v}$ dependency of $\sigma(v)$,

$$R(v) = \left(\frac{\rho N_A \sigma_0 v_0}{A_W}\right) n(v) \tag{3.26}$$

and hence the total thermal neutron reaction rate is

$$R = \int R(v)dv = \left(\frac{\rho N_A \sigma_0 v_0}{A_W}\right) \int n(v)dv = \frac{\rho N_A \sigma_0 v_0}{A_W} n_T \tag{3.27}$$

where n_T = total thermal neutron density. Equation (3.27) holds for the reaction rate of a $\frac{1}{v}$ absorber in a neutron field predominantly thermal. Multiplying and dividing Equation (3.27) by the average neutron velocity we have

$$R = \frac{\rho N_A \sigma_0}{A_w} \left(\frac{v_0}{\bar{v}}\right) n_T \bar{v} = \frac{\rho N_A \sigma_0}{A_w} \left(\frac{v_0}{\bar{v}}\right) \Phi \qquad (3.28)$$

In other words the 2200 m/sec cross section such as tabulated in Appendix A for thermal neutrons must be corrected by the factor

$$\frac{v_0}{\bar{v}} = \frac{\sqrt{\pi}}{2}$$

to derive the reaction rate from flux measurements or vice versa. Thus for thermal neutron reaction rate calculations, given the thermal neutron flux, the appropriate cross section to use is

$$\bar{\sigma} = \sigma_{293^\circ K}^{2200\ m/sec} \frac{\sqrt{\pi}}{2} \sqrt{\frac{293}{T}} \qquad (3.29)$$

No such corrections are required in the use of fast neutron cross sections. It is seen that the microscopic cross section is a property of each individual nucleus. Therefore, in a chemical compound, the total microscopic cross section is the sum of the constituent nuclei, i.e., for a compound

$$\sigma = \sum_i \gamma_i \sigma_i \qquad (3.30)$$

where γ_i is the number of atoms of the ith kind in the compound. For example, for UO_2,

$$\sigma = (1)\sigma_u + (2)\sigma_0$$

The macroscopic cross section for a compound is then

$$\Sigma = \frac{\rho_{compound} N_A \sigma}{M_{compound}} \qquad (3.31)$$

where $M_{compound}$ is the molecular weight of the compound and σ is given by Equation (3.30).

3.2 Frenkel Pair Formation

As mentioned earlier, fast neutrons are the major cause of radiation damage in a nuclear reactor environment. The fission event in a reactor forms fission fragments and neutrons. Both the fragments and fast neutrons are very energetic, being emitted from the fission sites at high velocities. If they are in a solid (moderator, fuel, structural component, etc.) the particles will smash through the solid lattice, dissipating energy by ionization, and transferring energy to atoms in the lattice by "billiard ball" type collisions. For neutrons the lighter the lattice atom, the more energy that is transferred. The energy transferred from neutron to lattice atom is often enough to eject the atom from its lattice site. This primary knock-on atom (PKA) may strike other atoms (secondaries), which in turn may be displaced from their lattice sites, striking other atoms (tertiaries) in turn. The sum of all primary, secondary, and tertiary atom displacements plus those generated by fission fragments represents the total displacements.

Neutrons may cover substantial distances while fission fragments and knock-on atoms are displaced a limited number of lattice positions. The displaced atoms leave a vacancy at the lattice site and may end up squeezed between lattice atoms in an interstitial position. Vacant lattice sites are designated as vacancies, and atoms forced between lattice positions are termed interstitials. Vacancies and interstitials may recombine (annealing), removing part of the damage. Finally a saturation condition is reached. The residual defects, whether simple or complex, represent a major share of damage occurring in a metal. Vacancy-interstitial pairs are often called Frenkel pairs, or Frenkel defects, and can cause changes in mechanical and physical properties. A schematic picture of neutron-induced radiation damage is shown in Figure 3.5 [Reference (3-6)].

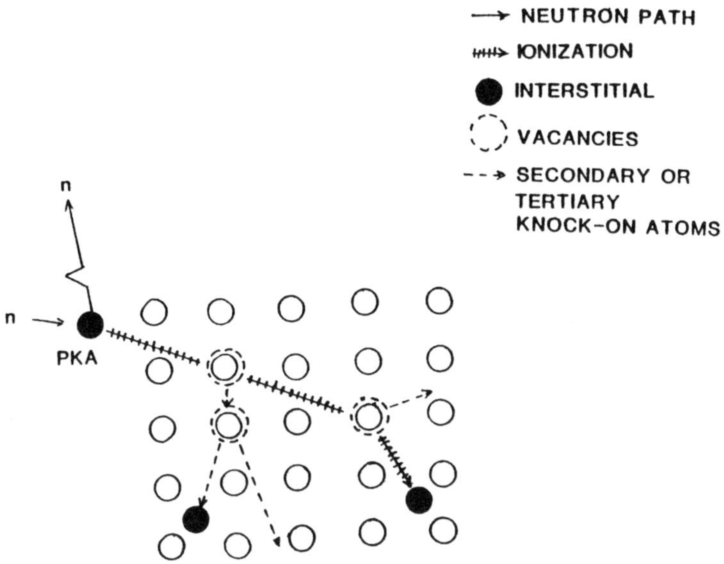

FIGURE 3.5 Schematic picture of neutron-induced radiation damage in metals.

In order to calculate how many Frenkel defects are formed, one must have a knowledge of a) cross section of neutron-nucleus interaction, b) reaction rate, and c) recombination (annealing) process.

It should be mentioned that other types of defects occur, such as point defects, cluster defects, dislocations, etc. For a brief and general discussion of these types, the reader is referred to Reference (3-7). However, the remainder of this section of the text will concentrate on Frenkel pair defects, since this is a major mechanism of lattice damage.

3.2.1 Simplified analytical model for Frenkel pair formation in metals. Before calculating the rate of formation of Frenkel pairs, some simplifying assumptions will be made. These are a part of the so-called Kinchin and Pease model [Reference (3-8)].

1. The displacement process is a sequence of two-body, hard-sphere interactions with an isotropic distribution of energy.

2. Collisions are presumed to be elastic below some ionization energy, E_i, of that material.

3. A sharp threshold energy, E_d, for displacement is presumed; a struck atom will be displaced if it receives a quantity of energy greater than E_d. Also the striking atom must remain in the vacated site if its retained energy is less than E_d.

In this model, the rate of formation of PKA's per unit volume is, as shown in a previous section

$$R = \Sigma_d \Phi \tag{3.32}$$

where Σ_d is the fast neutron macroscopic cross section for displacement of atoms which may be taken as the scattering cross section. According to the Kinchin-Pease model, the average number of displacements (Frenkel pairs) occurring per PKA, ν, is

$$\nu = \overline{E}_p / 2E_d \tag{3.33}$$

where E_d = displacement energy of an atom from its lattice site. A value of 25 eV for E_d is typically used, based on an average of several experiments. $\overline{E}_p$ is the average energy transferred from the incident neutron to the struck atom. Note that by this simplified model, half of the energy transferred to the PKA goes into forming displacements; the other half is spent in ionization. To evaluate Equation (3.33), one must calculate $\bar{E}_p$. To illustrate this calculation, consider first a head-on elastic collision between a neutron and a "target" (lattice) atom, assumed to be negligibly bound to its site. The situation is depicted in Figure 3.6 and gives the maximum possible energy transfer. Linear momentum is conserved in the collision, so that

$$m_n v_0 + 0 = -m_n v + m_T V \qquad (3.34)$$

Further, in an elastic collision, kinetic energy is conserved, so that

$$\frac{1}{2} m_n v_0^2 = \frac{1}{2} m_n v^2 + \frac{1}{2} m_T V^2 \qquad (3.35)$$

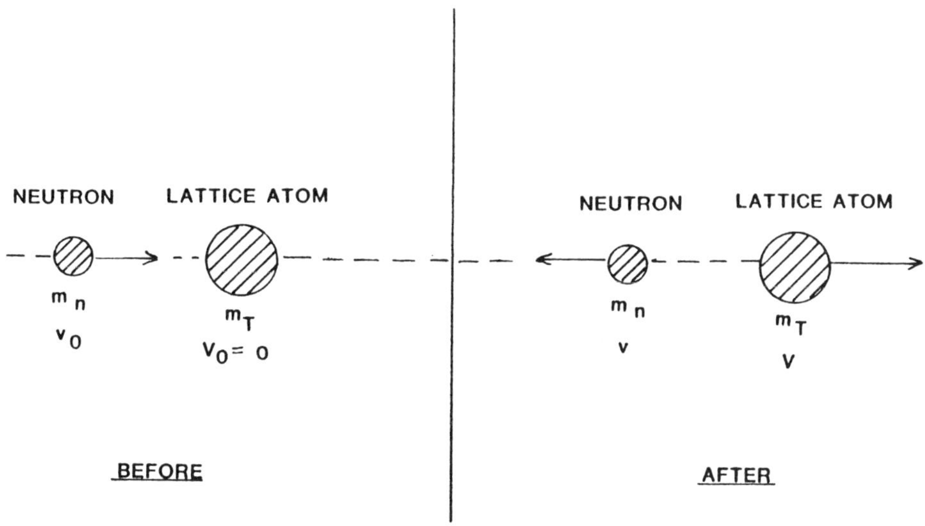

FIGURE 3.6 Head-on collision between neutron and lattice atom.

Solving Equations (3.34) and (3.35) simultaneously for the energy lost by the neutron in the collision (which is just the energy gained by the target atom E_p)

$$E_{p_{max}} = \frac{1}{2} m_n v_0^2 - \frac{1}{2} m_n v^2 = \frac{1}{2} m_T V^2$$

$$= \frac{4m_n m_T}{(m_n + m_T)^2} E_n \qquad (3.36)$$

where $E_n = \frac{1}{2} m_n v_0^2$ = initial energy of neutron. Now, in a "just grazing" collision, the neutron loses zero energy ($E_p = 0$), so that <u>on the average</u>

$$\overline{E}_p = \frac{1}{2} (E_{p_{max}} + 0) = \frac{2m_n m_T}{(m_n + m_T)^2} E_n \qquad (3.37)$$

[Equation (3.37) can be derived in a much more elaborate fashion, averaging collisions over all angles.]

<u>Illustration 3.4</u> Calculate the average energy of primary knock-ons, $\overline{E}_p$, for Zr, and Fe, when struck with 2-MeV neutrons. For Zr, the mass number A = 91 and for the neutron A = 1, so that

$$\overline{E}_p = \frac{(2)(91)}{(1+91)^2} (2 \text{ MeV}) = 0.043 \text{ MeV}$$

and for Fe, A = 56, so that

$$\overline{E}_p = \frac{(2)(56)}{(1+56)^2} (2 \text{ MeV}) = 0.069 \text{ MeV}$$

That is, for typical reactor materials, the primary knock-on atom acquires an energy ~50 KeV when struck by a neutron. Note that, from Equation (3.37), the lighter the target atom, the more energy it receives from the neutron in the average collision. This is the reason for using hydrogen (in the H_2O) as a moderator material in thermal power reactors.

One can now calculate the total number of Frenkel pairs formed per unit volume per unit time, L. From Equations (3.32), (3.33) and (3.36),

$$L = R\nu = \Sigma_s \Phi \nu = \Sigma_s \Phi \frac{\overline{E}_p}{2E_d} = \Sigma_s \Phi \frac{m_n m_T}{(m_n + m_T)^2} \frac{E_n}{E_d} \qquad (3.38)$$

Illustration 3.5 Calculate the Frenkel pair production rate, L, for Zr in a fast neutron flux of 1.87×10^{14} n/(cm^2)(sec). [This is the peak fast flux in the Zimmer reactor core.] Assume that the average energy required to displace an atom is $E_d = 25$ eV.

From Equation (3.38), $L = \Sigma_s \Phi \nu$. $\Sigma_s = \frac{1}{\lambda} = 0.263\ \text{cm}^{-1}$ (from Illustration 3.2).

$$\nu = \frac{\overline{E}_p}{2E_d} = \frac{0.043 \times 10^6}{(2)(25)} = 860 \text{ Frenkel pairs/PKA.}$$

Therefore

$$L = (0.263)(1.87 \times 10^{14})(860) = 4.23 \times 10^{16} \frac{\text{Frenkel pairs}}{(\text{cm}^3)(\text{sec})}$$

Note from this illustration that the number of secondary atoms displaced per PKA, in this case 860, completely dwarfs the initial event of the formation of the PKA, i.e. $\nu + 1 \cong \nu >> 1$. Hence the PKA, after being formed, does the lattice "damage," and not the neutron. The neutron generally has such a large mean free path that, after forming a PKA, it "disappears" from the vicinity of the damage.

3.2.2 Simplified analytical model for analysis of radiation effects in metals. In general, recombination (annealing) will occur during (and after) irradiation. The annealing rate is proportional to the square of the number of Frenkel pairs. Thus a very simple kinetic model to determine the density, n, of the Frenkel pairs, during irradiation is

$$\frac{dn}{dt} = L - An^2 \qquad (3.39)$$

where A is the recombination coefficient. The general solution to Equation (3.39) is

$$n(t) = \sqrt{\frac{L}{A}} \; \frac{(e^{2\sqrt{AL}\,t} - 1)}{(e^{2\sqrt{AL}\,t} + 1)} \qquad (3.40)$$

The saturation concentration of defects ($t = \infty$) is thus

$$n_\infty = \sqrt{\frac{L}{A}} \qquad (3.41)$$

When the source of the defect production is removed, i.e., when L = 0, the defect number density decays according to

$$n(t) = \frac{1}{\frac{1}{n(t_0)} + At} \qquad (3.42)$$

where t_0 is the time of irradiation.

Since the unirradiated equilibrium concentration n_{th} of defects is proportional to $\exp[-E_v/2kT]$ where E_v is the enthalpy of formation of a defect, it seems logical to assume a gross temperature dependency of the recombination coefficient of

$$A = A_0 e^{-E_v/kT} \qquad (3.43)$$

Data on Frenkel pair recombination coefficients for various metals are sparse. Recent calorimetric work with irradiated copper [Reference (3-9)] indicates a gross value of $A = 1.3 \times 10^{-27}$ cm^3/sec at temperatures below 300°C.

Illustration 3.6 Calculate the number of displaced atoms in zirconium cladding in a reactor, assuming absolutely no annealing, after 10 months of irradiation in a peak fast neutron flux of 1.87×10^{14} n/(cm^2)(sec). What fraction of the atoms are displaced?

From Illustration 3.5, the number of displaced atoms per unit volume per unit time is $L = 4.23 \times 10^{16}$. After 10 months of irradiation, the number of displaced atoms per unit volume is

$$4.23 \times 10^{16}(10)(30)(24)(3600) = 1.1 \times 10^{24}$$

The fraction of displaced atoms is thus $1.1 \times 10^{24}/N$, where N is the atomic density. Thus

$$\text{fraction of displaced atoms} = \frac{1.1 \times 10^{24}}{4.25 \times 10^{22}} = 25.8$$

(which is an absurd answer!)

<u>Illustration 3.7</u> Same as previous illustration, but now consider annealing effects. For the lack of better data, take the recombination coefficient to be that mentioned previously for copper, i.e., $A = 1.3 \times 10^{-27}$ cm^3/sec. Considering the defect concentration to be saturated, from Equation (3.41),

$$n_\infty = \sqrt{\frac{L}{A}} = \left(\frac{4.23 \times 10^{16}}{1.3 \times 10^{-27}}\right)^{1/2} = 5.7 \times 10^{21}\ \frac{\text{displaced atoms}}{(\text{cm}^3)}.$$

Thus the fraction of displaced atoms at equilibrium is

$$\frac{5.7 \times 10^{21}}{4.25 \times 10^{22}} = 0.13$$

Thus in this (rather extreme) example, in equilibrium, 13 percent of the atoms are displaced from their lattice sites! One would thus expect significant changes in the mechanical and thermal properties of this material due to the radiation damage. In fact, such calculations provide a crude measure for anticipated damage effects in metals. If the estimated fraction of defects is of the order of a percent or so, damage (changes in physical properties) may be expected. On the other hand, if the estimated fraction is much less than one percent, serious damage may not be anticipated. One of the major drawbacks currently with this estimate is the lack of data on the recombination coefficient.

3.2.3 Stored energy in Frenkel defects. From calorimetric measurements one can determine the stored energy ΔH in the irradiated metal due to the effects of the irradiation,

$$\Delta H = n(t)E_v \tag{3.44}$$

From Equation (3.42), this stored energy decays after irradiation as

$$\Delta H = \frac{1}{\frac{1}{\Delta H_0} + (\frac{A}{E_v})t_d} \tag{3.45}$$

where t_d is the decay time, and from Equation (3.40)

$$\Delta H_0 = n(t_0)E_v = E_v\sqrt{\frac{L}{A}}\,\frac{(e^{2\sqrt{AL}\,t_0}-1)}{(e^{2\sqrt{AL}\,t_0}+1)} \tag{3.46}$$

where t_0 is the time of irradiation.

For small irradiation times, Equation (3.46) becomes

$$\Delta H_0 \cong E_v L t_0 \tag{3.47}$$

By annealing at various temperatures after irradiation until measurement of ΔH, by calorimetric methods one can determine E_v from Equation (3.47) [(through Equation 3.45)]. By a fit to the data, one also thus obtains A_0.

Belwitt, Coltman, Holmes, and Noggle irradiated massive samples of Cu and Al with fast neutron fluence up to 4×10^{17} n/cm^2 and measured 0.3 cal/mole in the range of 20-60°K [Reference (3-10)]. Kinchin and Thompson have measured the stored energy in molybdenum after neutron irradiation up to 1.3×10^{19} n/cm^2 at 300°K. The stored energy measured was 0.4 cal/gm up to 525°K. There was found a close resemblance of annealing between high temperature resistivity change and the stored energy release [Reference (3-11)]. Irradiation of copper by 12-MeV deuterons was performed by Overhauser and the stored energy was measured up to 0.052 cal/gm per 10^{17} ions/cm^2 in the range of 173°K to 293°K [Reference (3-12)]. Nilan and Granoto also measured the release of stored energy of 0.75 cal/g per 10^{17} deuterons/cm^2 during annealing from 15-80°K of Cu irradiated with 11-MeV deuterons [Reference (3-13)]. Recently, Anno, Pool, and Wooten of the University of Cincinnati conducted a series of experiments with copper foils [Reference (3-14)]. Irradiation with 1-MeV deuterons for only 40 minutes at low beam current of 40 microamperes produced stored energy of 730 cal/gm-atom. With crude estimates, this is equivalent to about 0.8 eV of stored energy per defect, which is higher than previously expected. The data also predicted that the energy stored in a material at saturation due to radiation effects can be many times the unirradiated heat of solution. One therefore suspects that such material as the cladding in a reactor which have been exposed to a high fast neutron fluence will have an appreciable amount of stored energy.

Illustration 3.8 Calculate the stored energy in Zr at equilibrium defect concentration in a fast flux of 1.87×10^{14} n/(cm^2)(sec). Assume $E_v = 0.8$ eV where E_v = stored energy per defect. The stored energy per unit volume at equilibrium is

$$= n \times E_v$$
$$= 5.7 \times 10^{21} \times 0.8 = 4.6 \times 10^{21} \text{ eV/cm}^3$$
$$= \frac{(4.6 \times 10^{21})(1.6 \times 10^{-19})(10^3)}{(6.44)} = 1.1 \times 10^5 \text{ joules/kg}$$

3.3 Activation and Transmutation

One of the simplest forms of neutron radiation effects on materials is the absorption by a nucleus of a neutron to transmute it to another isotope, often radioactive. This simple effect leads both to induced radioactivity and to the introduction of a "foreign" atom species into the material.

3.3.1 Neutron activation. Let us assume that the immediate "daughter" product of a neutron absorption is radioactive. Such a reaction is typified by the (n,γ) reaction with gold, discussed in Section 3.1.2, in which stable ^{197}Au is transformed into radioactive ^{198}Au by the absorption of a neutron. The number of radioactive atoms N' formed per unit time, produced in a chunk of material of volume V, is given as

$$\frac{dN'}{dt} = RV = \Sigma_{act}\Phi V \tag{3.48}$$

where Σ_{act} is the macroscopic activation cross section, distinguished from the macroscopic absorption cross section only if every absorption does not lead to an activation (otherwise and usually, they are the same). Since the new species N' is radioactive, it will also decay during activation according to the exponential law of decay $e^{-\lambda t}$ where λ is the decay constant (related to the half-life $T_{1/2}$ by $\lambda = \frac{\ln 2}{T_{1/2}}$).

Thus the rate of change of the number of activated atoms (neglecting burnup of the activated species by further absorption) is

$$\frac{dN'}{dt} = RV - \lambda N' \tag{3.49}$$

It is convenient to introduce the concept of "activity" A of a material, since this quantity is the physically measured property,

$$A \equiv \lambda N' \quad \text{disintegrations/sec} \tag{3.50}$$

Rewriting Equation (3.49) in terms of activity

$$\frac{dA}{dt} = \lambda RV - \lambda A \tag{3.51}$$

Since A(0) = 0 can be assumed, the solution of Equation (3.51) is

$$A = RV(1-e^{-\lambda t}) = \Sigma_{act}\Phi V(1-e^{-\lambda t}) \tag{3.52}$$

where t is the activation (exposure) time. At the end of the irradiation, the chunk of material will just decay exponentially, and hence at any time T after the exposure

$$A = \Sigma_{act}\Phi V(1-e^{-\lambda t})\, e^{-\lambda T} \tag{3.53}$$

The unit of activity is the curie, where 1 curie = 3.70×10^{10} disintegrations/sec. (This particular value was earlier established as the number of disintegrations per second of one gram of radium.) Since most materials exposed in a reactor have a thermal neutron absorption cross section much greater than the fast neutron absorption cross section (see for example, Figure 3.4), most induced radioactivity is due to thermal neutrons. However, for fusion reactors, where fast neutrons predominate, activation of reaction chamber components is also a problem.

3.3.2 Effects of transmutations. The number of transmuted atoms present per unit volume, n', for the simple formation and decay scheme presented in the previous section is, from Equation (3.52),

$$n' = \frac{N'}{V} = \frac{\Sigma_{act}\,\Phi}{\lambda}\,(1-e^{-\lambda t}) \qquad (3.54)$$

These new species are "impurity" atoms in a material, and, in fact, can introduce changes in physical properties. However, for metals the changes are usually quite small, if not negligible.

Illustration 3.9 Calculate the fraction of transmuted atoms formed in the (mostly iron) pressure vessel in a thermal reactor after continuous exposure to a thermal neutron flux of 10^{10} n/(cm^2)(sec) for 10 years. (Assume thermal neutron temperature is 200 °C.)

The half-life of ^{59}Fe, formed from the 0.31 percent abundant ^{58}Fe isotope, is 45 days. Thus after 10 years the material will be saturated and Equation (3.54) becomes

$$n' = \frac{\Sigma_{act}\Phi}{\lambda}$$

where $\Sigma_{act} = \dfrac{\rho N_A}{A_w}\,\sigma_{act}\gamma$

where γ is the abundance fraction. $(\lambda = \frac{\ln 2}{45} = 0.0154\ \text{day}^{-1} = 1.78 \times 10^{-7}\ \text{sec}^{-1})$. However the number density of iron, itself, is

$$N = \frac{\rho N_A}{A_w}$$

So that the fraction of transmuted atoms (from the source used in this calcuation) is

$$\frac{n'}{N} = \frac{\Phi\sigma_{act}\gamma}{\lambda} = \frac{(10^{10})(1.2 \times 10^{-24})}{(1.78 \times 10^{-7})}\frac{\sqrt{\pi}}{2}\sqrt{\frac{293}{473}}\,(0.0031)$$

$$= 1.46 \times 10^{-10}$$

This illustration indicates that the general magnitude of impurity atoms introduced by neutron transmutations is indeed small, so that little effects are to be expected generally from this source of radiation effects.

There is, however, another effect of transmutation, often neglected. This is the affect on the cross section of the material due to the new nuclear species introduced by the transmutations. The cross sections for the natural elements have been rather extensively measured, as demonstrated by the information presented in Appendix A and the references from which it was extracted. However, these and similar compilations of cross sections also indicate the need for data on the cross sections of radioactive nuclei which are produced by transmutation due to neutron reactions with the natural elements. For example, the thermal absorption or activation cross section is listed for only 338 radioactive nuclei out of a possible 1,785 known radioactive nuclides. In general, the accuracy to which even this small group of cross sections is known is poor and 21 of the 338 are just known to be "greater than" or "less than" certain values.

A knowledge of the thermal neutron absorption cross sections of many of the radioactive nuclei has been important to several facets of reactor development. The formation of daughter products with significant cross section can influence choice of reactor structural materials, reactor control rod materials and materials used for experiments with the reactor radiations. A classic example of the importance of the daughter product cross sections in choice of reactor materials is the use of tantalum as a stabilizer in stainless steel. It was originally thought that tantalum was very undesirable in this application because of the radioactivity resulting from the formation of the long-lived tantalum-182 isotope during irradiation [Reference (3-15)]. However, it was later discovered that tantalum-182 has such a large thermal neutron absorption cross section (approximately 17,000 barns) that it is rapidly burned out to short-lived isotopes which decay to stable tungsten isotopes with low cross sections [Reference (3-16)]. An example of the importance of the cross sections of radioactive nuclei to selection of control rod materials is the breeding of high cross section isotopes from the burnup of natural europium which prevents rapid loss of control due to burnup while in the reactor and hence increases the lifetime of control rods made from this material [Reference (3-17)].

Measurement of the cross sections of the individual radioactive isotopes is made difficult by the limited availability of these materials (generally small yields during irradiation) with the added problems associated with their separation from the parent isotopes, their decay and associated radioactivity. The effective absorption cross section of a material of mass number A under irradiation changes primarily due to neutron absorption which forms a daughter product of the same isotope but with the mass number increased to A + 1, and with a different cross section than the parent. Neutron capture by the daughter product can then form a second isotope of the material with mass number A + 2. Successive neutron captures lead to isotopes of higher mass numbers but in increasingly smaller yields. Analytically. the process may be described as follows. If ${}^{A}X_{Z}$ is the concentration of the initial isotope, where Z is the atomic number and A the mass number, then the rate of depletion of the original isotope under irradiation is

$$\frac{d}{dt} ({}^{A}X_{Z}) = - \sigma_{Z,A} \Phi {}^{A}X_{Z} \tag{3.55}$$

where

$\sigma_{Z,A}$ = microscopic thermal absorption cross section of isotope with atomic number Z and mass number A, cm^2

Φ = thermal neutron flux, $n/(cm^2)(sec)$

Accordingly, the concentration of the daughter product is given as

$$\frac{d}{dt}\left({}^{A+1}X_Z\right) = \sigma_{Z,A}\Phi\,{}^{A}X_Z - \lambda_{Z,A+1}\,{}^{A+1}X_Z - \sigma_{Z,A+1}\Phi\,{}^{A+1}X_Z \tag{3.56}$$

where

$\lambda_{Z,A+1}$ = decay constant of the daughter product, sec^{-1} ($\lambda_{Z,A+1} = 0$ if daughter is stable)

In general, the concentration of the nth daughter product formed by successive neutron capture is given by

$$\frac{d}{dt}\left({}^{A+n}X_Z\right) = \sigma_{Z,A+n-1}\Phi\,{}^{A+n-1}X_Z - \lambda_{Z,A+n}\,{}^{A+n}X_Z - \sigma_{Z,A+n}\Phi\,{}^{A+n}X_Z \tag{3.57}$$

Considering only isotopes produced in this manner, the effective cross section is given as

$$\overline{\sigma} = \frac{\sum_{i=0}^{n} {}^{A+i}X_Z\,\sigma_{Z,A+i}}{X_0} \tag{3.58}$$

where

X_0 = initial concentration of the parent isotope.

However, the daughter product formed at each neutron capture is generally radioactive and may decay with the emission of a β^- particle to an element of the same mass number but with the atomic number increased by one unit. Hence, another species will be present and contribute to the effective cross section. The general differential equations governing the concentrations of isotopes formed by β^- decay of the daughter product is

$$\frac{d}{dt}({}^{A+n}X_{Z+1}) = \lambda_{Z,A+n}\,{}^{A+n}X_Z - \sigma_{Z+1,A+n}\Phi\,{}^{A+n}X_{Z+1} - \lambda_{Z+1,A+n}\,{}^{A+n}X_{Z+1} \tag{3.59}$$

Thus, the effective cross section must be "corrected" to include this species.

$$\bar{\sigma} = \sum_{i=0}^{n} \left({}^{A+i}X_Z\,\sigma_{Z,A+i} + {}^{A+i}X_{Z+1}\,\sigma_{Z+1,A+i}\right) \tag{3.60}$$

Higher order reactions (β^- decay of the ${}^{A+n}X_{Z+1}$ isotope, etc.) and modes of decay of the daughter products other than by emission of β^- particles must also be included in a general analysis, although their contributions to the effective cross section will, in general, be small.

From the foregoing discussion, it can be seen that a knowledge of the absorption cross section of many radioactive isotopes is required to accurately predict the effective absorption cross section of a material as a function of irradiation. Only in special cases, where all products but ${}^{A}X_Z$ and ${}^{A+1}X_Z$ can be neglected and when $\sigma_{Z,A+1}$ is known, does the problem reduce to a simple one. Thus, to predict the cross section behavior with irradiation, one must either know the absorption cross sections of all isotopes involved or use experimental determinations of the behavior.

For simplicity, consider as an example the time behavior of the effective absorption cross section of irradiated gold. Upon the absorption of a neutron the parent isotope, gold-197, with an absorption cross section of 98.0 barns, is transmuted to gold-198, with an absorption cross section of 25,800 barns. Gold-198 beta decays to stable mercury-198 with a half-life of 2.7 days. The cross section of mercury-198 is very small (approximately 18 mb). However, some of the gold-198 is transmuted to gold-199, etc. In calculations of the effective cross section of irradiated gold, only those isotopes shown in Figure 3.7 were considered. The results of the calculations are shown in Figure 3.8 for various constant thermal neutron flux levels. This example, although it may not be typical, points out that the effective cross section can be significantly different from the "cold, clean" value. A more typical material is stainless steel. The variation in cross section with time at various thermal neutron fluxes is shown in Figure 3.9. The composition assumed for this material was 69.1% Fe, 18% Cr, 11% Ni, 1% Si, 0.7% Mn, and 0.2% C. Because of such cross section changes, one questions the use of three-significant-figure values of cross sections in computer codes calculating properties of "aged" reactors.

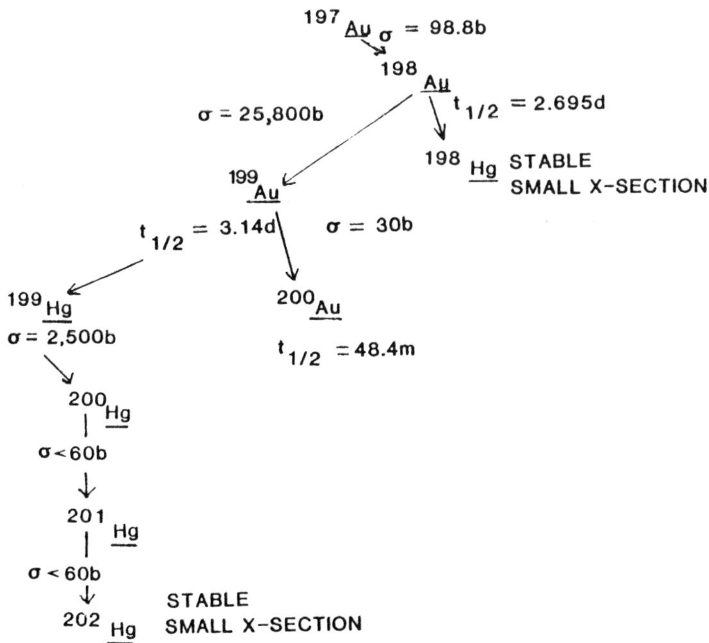

FIGURE 3.7 Isotopes considered in calculating an effective cross section for irradiated gold.

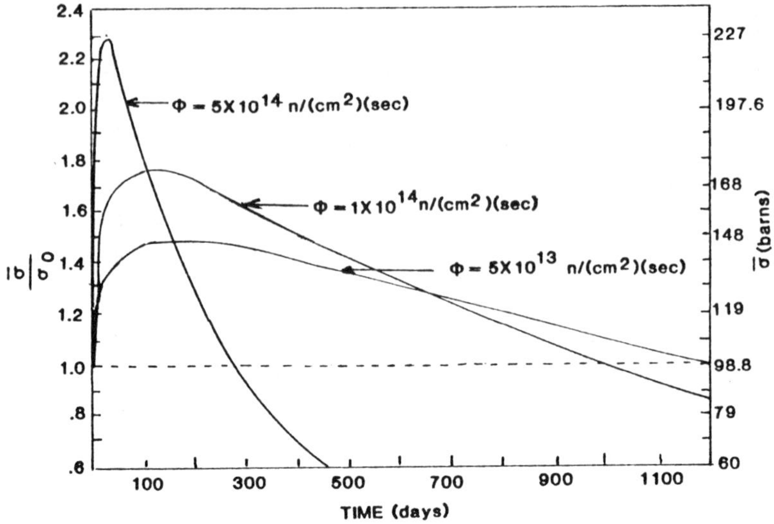

FIGURE 3.8 Change in the effective cross section of gold as a function of irradiation time and thermal neutron flux.

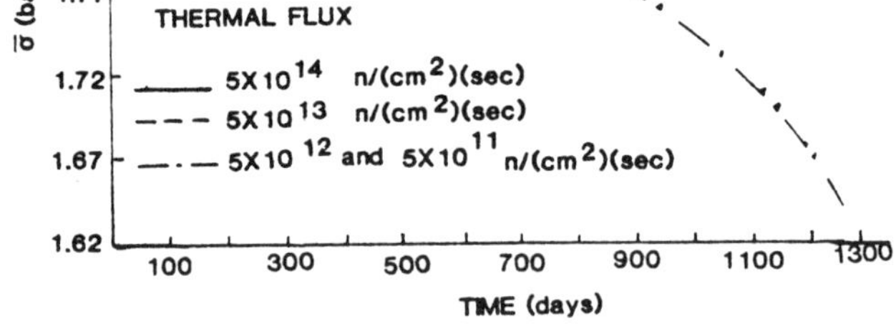

FIGURE 3.9 Change in the effective cross section of stainless steel as a function of irradiation time and thermal neutron flux.

REFERENCES

(3-1) Billington, D.S. and Crawford, Jr., J.H., Radiation Damage in Solids, Princeton U.P., Princeton, New Jersey, 1961.

(3-2) Hughes, D.J. and Harvey, J.A., "Neutron Cross Sections," BNL-325 early report series.

(3-3) Garber, D.I. and Kinsey, R.R., "Neutron Cross Sections," BNL-325, Edition 3, Vol. 2, 1976.

(3-4) Hughes, D.J., Pile Neutron Research, Addison-Wesley Publishing Co., Inc. 1953.

(3-5) Kaplan, Irving, Nuclear Physics, Addison-Wesley Publishing Co., Inc., 1955.

(3-6) Bush, Spencer H., Irradiation Effects in Cladding and Structural Materials, Rowman and Littlefield, Inc., 1965.

(3-7) Vook, Frederick L., "Effects of Radiation on Materials," Phys. Today, 34-40, September, 1975.

(3-8) Kinchin, G.H. and Pease, R.S., Repts. Prog. in Phys. 18, 1, 1955.

(3-9) Wooten, A.L., "Determining the Stored Energy in Irradiated Copper," M.S. Thesis, University of Cincinnati, 1981.

(3-10) Blewitt, T.H., Coltman, R.R., Holmes, D.K., and Noggle, T.S., Dislocations and Mechanical Properties, John Wiley, New York, 1958, p. 603.

(3-11) Kinchin, G.H. and Thompson, M.W., Jour. Nuclear Energy, 6, 275, 1958.

(3-12) Overhauser, A.W., Phys. Rev., 94, 1551, 1954.

(3-13) Granato, A.V., and Nilan, T.G., Phys. Rev., 137, A1250, 1965.

(3-14) Anno, J.N., Pool, M.J., and Wooten, A.L., University of Cincinnati, unpublished data, 1980.

(3-15) AECD-3647, The Reactor Handbook, Vol. 3, "General Properties of Materials," p. 266.

(3-16) P/590, International Conference on the Peaceful Uses of Atomic Energy, (First Geneva Conference), "Status of Information on Reactor Material Cross Sections," by H. Hurwitz, Jr., May, 1955.

(3-17) Stevens, H.E., "Nuclear Requirements for Control Materials," Nuclear Science and Engineering, Vol. 4, No. 3, 376, September, 1958.

PROBLEMS FOR CHAPTER 3

1. A typical average thermal neutron flux in a power reactor is 5×10^{13} n/(cm^2)(sec). A typical macroscopic fission cross section is 0.05 cm^{-1}. What is the power density for such a system in units of watts/cm^3, assuming 194 MeV/fission?

2. What is the mean free path of thermal neutrons between fission for the reactor in problem 1? Compare this with the size of a 1000 megawatt (thermal) reactor, assuming it to be a cube with the power density given from problem 1, i.e. qualitatively would you expect many neutrons to escape out the sides of the reactor?

3. The average fast neutron flux (>0.1 MeV) in a power reactor core is given as 10^{14} n/(cm^2)(sec). Assuming most of the "important" material in the reactor core is water, what is the mean free path of these neutrons for a scattering reaction? for an absorption reaction?

4. What is the absorption cross section for a "$\frac{1}{v}$" absorber at 2.53 eV neutron energy at 293°K, given σ_o = 100 barns at 2200 m/sec and 293°K?

5. Derive the expression for the (maximum) energy gained by a target atom (PKA) upon a head-on elastic collision with a neutron.

6. A 2-MeV neutron has a head-on collision with an iron atom (mass = 57 amu) in a piece of iron. If the displacement energy is E_d = 25 eV for dislodging an iron atom from its lattice site, how many displacements will occur on the average?

7. When Frenkel pair production is in equilibrium with recombination, how does the equilibrium concentration of Frenkel pairs vary with neutron flux?

8. A gold foil of mass 50 mg is irradiated in a thermal neutron flux of 5 x 10^{11}n/(cm^2)(sec) for one day. What is its activity (in curies) at the end of the irradiation? after 10 days of decay?

4 INTERACTIONS OF GAMMA RAYS WITH MATERIALS

In the previous chapter, the basic interactions of neutrons with material were discussed. This present chapter presents the interactions of the other most important type of radiation to nuclear engineers, namely gamma rays.

4.1 What is a Gamma Ray?

The photons, or electromagnetic waves, emitted by nuclei range in energy up to several MeV, and are traditionally called gamma rays. The dual nature of a gamma ray makes us able to treat it either as an electromagnetic wave propagating with the speed of light in vacuum, or as a photon emitted discontinuously as a little burst of energy E which is proportional to ν, the frequency. That is

$$E = h\nu \tag{4.1}$$

where h = Planck's constant = 6.63×10^{-34} joules/sec. According to the famous relationship Einstein obtained from postulates of special relativity concerning mass and energy

$$E = mc^2 = T + m_0c^2 \tag{4.2}$$

where E is the total energy of the body and T is the kinetic energy, and m_0 is the rest mass. It is useful to note that the conversion factor between the unit of mass (amu) and the unit of energy (MeV) is 931.48, i.e.

$$1 \text{ amu} = 931.48 \text{ MeV}$$

A gamma ray, which has a zero rest mass, thus has an effective mass equivalent to

$$"m" = \frac{E}{c^2} = \frac{h\nu}{c^2} \qquad (4.3)$$

<u>Illustration 4.1</u> Calculate the frequency and wavelength of a 1-MeV gamma photon.

From Equation (4.1)

$$E = h\nu = h\frac{c}{\lambda}$$

Therefore

$$\nu = \frac{E}{h} = \frac{1\text{MeV} \times 1.6 \times 10^{-6}\text{erg/MeV}}{6.626 \times 10^{-27} \text{ erg sec}}$$

$$= 2.42 \times 10^{20}\text{sec}^{-1}$$

$$\lambda = \frac{c}{\nu} = \frac{3 \times 10^{10} \text{ cm/sec}}{2.42 \times 10^{20}\text{sec}^{-1}} = 1.24 \times 10^{-10}\text{cm}$$

<u>Illustration 4.2</u> Calculate the effective mass of a 1-MeV photon.

From Equation (4.3)

$$"m" = E/c^2$$

$$= \frac{1 \text{ MeV} \times 1.6 \times 10^{-6} \text{ erg/MeV}}{(3 \times 10^{10} \text{ cm/sec})^2}$$

$$- 1.78 \times 10^{-27}\text{g}$$

or = 1.95 m_e (m_e= electron rest mass)

4.2 Sources of Gamma Rays in a Reactor

There are three primary gamma ray sources of interest to a nuclear power plant:

a. Capture gamma ray. In a radiative capture reaction, the excited compound nucleus emits its excess energy as gamma radiation. This process can be exemplified by the reaction

$$ {}^{1}_{0}n + {}^{58}_{26}Fe \longrightarrow {}^{59}_{26}Fe^{*} \longrightarrow {}^{59}_{26}Fe + \gamma \qquad (4.4)$$

b. Decay gamma ray. This kind of gamma ray comes mostly from the fission fragments and activation of structural materials. For example

$$ {}^{59}_{26}Fe \longrightarrow {}^{0}_{-1}\beta + {}^{59}_{27}Co + \gamma \qquad (4.5)$$

c. Internal excitation of fission fragments at birth. The excited fission fragments, immediately after the fission, give off gamma rays to de-excite, then go to various decay channels.

Am empirical fit to the data from various sources gives the source spectrum for all gamma rays resulting from one fission as

$$ N(E) = 14e^{-1.10E},\ MeV^{-1} \qquad (4.6)$$

where E is the gamma energy in MeV. The formula applies over the range of approximately 0.5-7 MeV. The average photon energy is about 1 MeV. Capture gammas (from thermal neutron capture) add to this spectrum, but in a manner dependent upon core materials, and therefore difficult to estimate.

4.3 The Process Involved in the Interaction of Gamma Rays with Materials

The interactions of high-energy photons with materials have received intensive study since the discovery of the X-ray, with a relatively recent impetus provided by reactor shielding problems [Reference (4-1) through (4-5)]. A brief review of these interactions and their probability provides technical background for this discussion. The three most important processes are the Compton Effect, Photoelectric Effect, and Pair Production.

<u>4.3.1 Compton effect</u>. In the Compton process, a photon of initial energy $h\nu_0$ strikes an orbital electron of an atom, ejecting the electron at angle Φ with the original path of the photon. The scattered photon is reduced in energy to $h\nu'$ and leaves the collision site at angle θ with the original path. The process is illustrated in Figure 4.1(a).

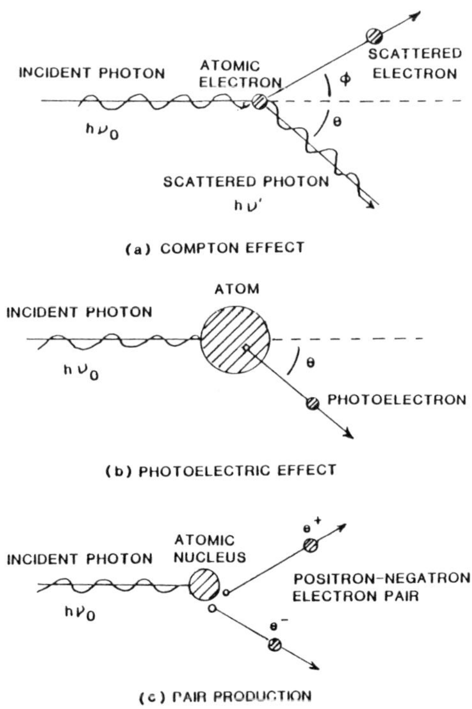

FIGURE 4.1 Principal interaction of photons with materials.

The maximum energy transferred to the electron is

$$E_{max} = \frac{h\nu_0}{1 + \frac{1}{2\alpha}} \tag{4.7}$$

where

$$\alpha = \frac{h\nu_0}{m_0c^2}$$

Since the rest-mass electron energy $m_0c^2 = 0.51$ MeV, gamma photons in the MeV range may transfer nearly all of their energy to the ejected electron. The <u>average</u> fractional energy lost by the photon in a Compton encounter is shown graphically in Figure 4.2. As seen from this graph, photons with energy greater than 1 MeV lose on the average half or more of their energy in a single event. Photons with initial energy in the range of 10 KeV lose only a few percent of their energy in a Compton event.

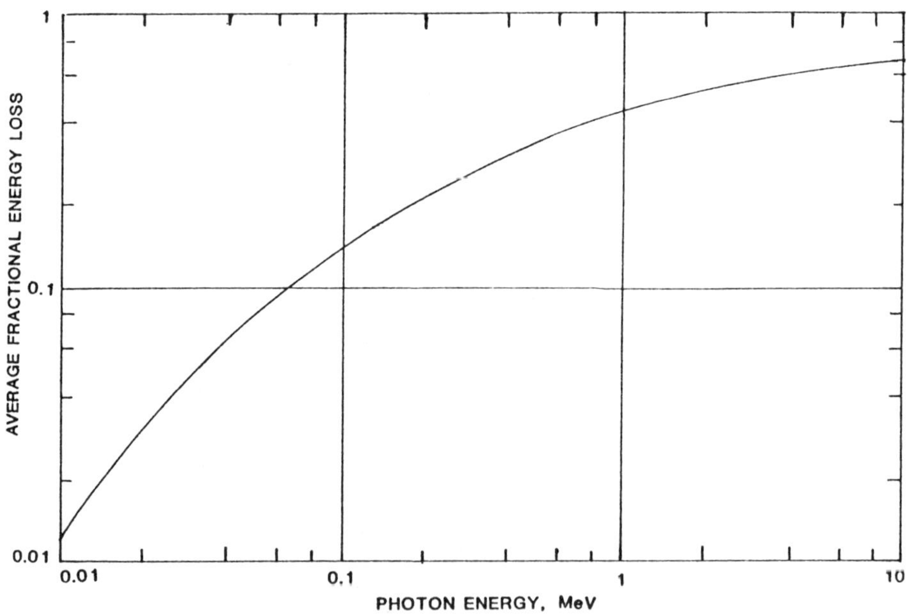

FIGURE 4.2 Average fractional energy lost in a Compton scattering event versus incident photon energy.

The average cross section for Compton scattering is

$$\sigma_c = Z\sigma_e \text{ cm}^2/\text{atom} \tag{4.8}$$

where Z is the atomic number of the scattering atom and σ_e is the collision cross section per electron, which decreases with photon energy from about 2/3 barn for very-low-energy photons to 0.21 barn for a 1-MeV photon, i.e., σ_e is not a strong function of photon energy. Equation (4.8) shows that the Compton cross section is directly proportional to the atomic number of the material.

4.3.2 Photoelectric effect. In the photoelectric effect, the incoming photon, $h\nu_0$, is completely absorbed and an electron (usually from the K- or L-shell) is ejected from the atom with energy

$$E = h\nu_0 - B_e \tag{4.9}$$

where B_e is the binding energy of the ejected electron (see Figure 4.1(b)). About 80 percent of the photoelectrons come from the K-shell, provided the incident photon energy exceeds the K-shell binding energy. The remainder of the energy (the binding energy) appears as characteristic X-rays or Auger electrons. For high-Z material, most of this energy appears as X-rays. The fraction of the binding energy appearing as X-rays, i.e., the radiative yield, for the K-shell is

$$\omega_K = \frac{1}{1 + \frac{1.12 \times 10^6}{Z^4}} \tag{4.10}$$

For example, for lead (Pb), $\omega_K = 0.975$.

The cross section for the photoelectric effect is given for K-shell electrons approximately as

$$\sigma_K = \frac{CZ^5}{(h\nu_0)^{7/2}} \text{ cm}^2/\text{atom} \qquad (4.11)$$

where C is a constant, about 10^{-33} when $h\nu_0$ is expressed in MeV. This relation shows that the cross section is a very strong function of atomic number Z. (Actually, the exponent of Z is a function of photon energy and varies from about 4 for 0.1-MeV photons to near 5 for 10-MeV photons.) Also, the cross section becomes larger for lower energy photons. As an example, for a high-Z material such as uranium, at gamma energies near the K-shell binding energy (0.16 MeV), the photoelectric-effect cross section is about 10^4 barns/atom compared to the Compton cross section under these same circumstances of 50 barns/atom. However, for reactor gammas, the Compton effect is usually the larger one.

4.3.3 Pair production. For photon energies greater than 1.02 MeV, a photon in the presence of an atomic nucleus may be transformed into a positron-negatron electron pair, illustrated in Figure 4.1(c). Part of the energy of the photon transforms to mass of the electrons and the remainder appears as kinetic energy, i.e.,

$$h\nu_0 = (E^- + m_0c^2) + (E^+ + m_0c^2) \qquad (4.12)$$

Very soon after its production, the positron is slowed down by collisions with atoms until it is practically at rest. It then interacts with an electron to form two annihilation photons, each with 0.511 MeV.

For photon energy of several MeV, the cross section for pair production is

$$\sigma_p = \sigma_0 Z^2 \left[\frac{28}{9} \ln \left(\frac{2h\nu_0}{m_0c^2}\right) - \frac{218}{27}\right] \qquad (4.13)$$

where

$$\sigma_0 = \frac{1}{137} \left(\frac{e^2}{m_0 c^2}\right)^2 = 5.80 \times 10^{-28} \text{ cm}^2/\text{nucleus}$$

As seen from Equation (4.13), the cross section depends on the square of the atomic number and increases logarithmically with photon energy. For example, for a 2-MeV photon the pair-production cross section is 1.25 barns/atom for lead (Pb).

4.4 Gamma Heating

One of the simplest effects of gamma radiation on materials is that of thermal heating due to absorption. By the very definition of the rad unit as the absorption of 100 ergs/gm, gamma heating is given as (for any material)

$$0.00278 \frac{\text{watts}}{\text{gm}} \text{ per megarad/hr.}$$

However, for practical applications, one generally does not know precisely the rad dose rate to a particular material, but rather the dose rate (usually air equivalent) as determined by an external radiation detector (see Chapter 12). Thus it is generally necessary to determine the gamma heating from a dose rate expressed, for example, in units of Roentgen/hr. The remainder of this section of Chapter 4 presents a semiempirical method for so doing.

4.4.1 Gamma heating "theory." As seen from the preceding discussion, the total cross section for the reaction of gamma rays with a material of atomic number Z can be written approximately as

$$\sigma = \underbrace{a_1'Z}_{\text{Compton}} + \underbrace{a_2'Z^2}_{\text{Pair Production}} + \underbrace{a_3'Z^5}_{\text{Photoelectric Effect}} \tag{4.14}$$

where the a_i's depend on the gamma ray energy spectrum. One then assumes that on a per atom basis, for unit gamma dose rate, and with negligible self-shielding in the material, the heating is given by

$$G_0 = a_1 Z + a_2 Z^2 + a_3 Z^5 \quad \frac{\text{watts}}{\text{atom(Roentgen/hr)}} \tag{4.15}$$

where the a_i's must be determined empirically for a given energy spectrum, such as that in a reactor (and location within the reactor).

4.4.2 Gamma heating experiment results. Measurements were made on graphite, aluminum, stainless steel, iron, nickel, copper, cadmium, tin, and lead samples exposed to gamma radiation in and around a thermal reactor [Reference (4-6)]. The results are shown in Table 4.1 along with a description of the samples. The data are shown graphically in Figure 4.3. To determine the self-shielding in the samples, the heating in various diameter samples of lead was measured. The results of these measurements are tabulated in Table 4.2. These data indicate that self-shielding was appreciable for the 3/8-in. diameter lead sample. To determine the effect of variation in gamma energy spectrum on the heating, the stainless steel sample was heated at various positions in and around the core.

As seen from Figure 4.3, plotting the atomic number of the material versus heating per atom yields a smooth curve. These values are for thick samples. The value for lead at zero thickness taken from extrapolation of the data in Table 4.2 is 27.4 x 10^{-28} milliwatts per atom per Roentgen/hr. Also, the value found for graphite may be assumed to be a zero thickness value. If a third value for zero thickness heating is picked, the theory of gamma heating may be applied to find the constant a_i in Equation (4.15). By requiring that the corrections to the measured values be a constantly increasing percentage with atomic number, this third value may be picked systematically by trial and error. Values of a_i found in this way are

TABLE 4.1

Gamma Heating Data for Thick Samples[a]

Material	Atomic Number Z	Sample Wt. (grams)	Dimensions (Diameter) x (Length) inches	Gamma Dose Rate[b] 10^7 R/hr	Gamma Heating ($\frac{\text{milliwatts}}{\text{atom R/hr}}$)
Graphite	6	1.66	0.376 x 0.547	2.10	0.91×10^{-28}
Aluminum	13	2.37	0.376 x 0.474	2.46	2.04×10^{-28}
Stainless Steel	25.8	7.54	0.376 x 0.548	2.13	2.86×10^{-28}
Iron	26	6.64	0.377 x 0.521	2.79	2.88×10^{-28}
Nickel	28	8.15	0.376 x 0.522	2.76	3.22×10^{-28}
Copper	29	0.58	0.375 x 0.600	2.84	2.11×10^{-28}
Cadmium	49	6.95	0.380 x 0.522	2.85	6.88×10^{-28}
Tin	50	6.65	0.377 x 0.509	1.98	7.88×10^{-28}
Lead	82	10.25	0.376 x 0.505	2.36	1.68×10^{-27}

(a) In water, approximately four inches from a pool-type research reactor core face.

(b) Gamma dose rate determined by a CO_2-filled carbon-walled ionization chamber.

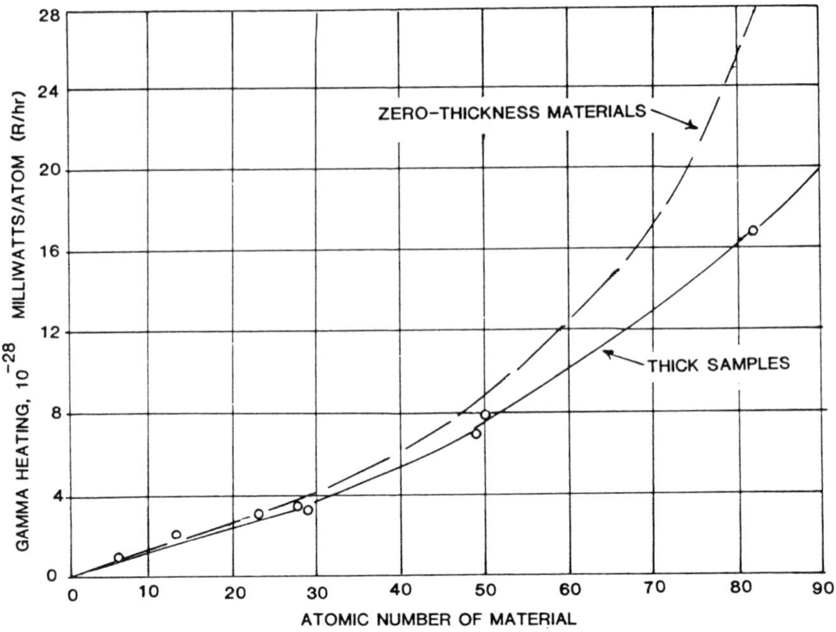

FIGURE 4.3 Gamma heating measurements with thick samples near a reactor core face.

[Reproduced with permission from J.N. Anno, "Factors Influencing the Effects of Reactor Radiations on Electrical Components," Nuclear Applications, 2, pp. 371-378, October, 1966]

TABLE 4.2

Gamma Heating Data for Varying Sizes of Lead Samples[a]

Sample Wt. (grams)	Dimensions (Diameter) x (Length) inches	Gamma Dose Rate (R/hr)	Gamma Heating (milliwatts / atom R/hr)
10.250	0.376 x 0.505	2.36×10^7	1.68×10^{-27}
4.609	0.251 x 0.522	2.63×10^7	1.80×10^{-27}
0.982	0.122 x 0.523	2.62×10^7	2.12×10^{-27}
0.279	0.065 x 0.526	2.70×10^7	2.47×10^{-27}

(a) In water, approximately four inches from the face of a pool-type research reactor core face.

$a_1 = 1.13 \times 10^{-29}$, $a_2 = 5.62 \times 10^{-32}$, and

$a_3 = 3.87 \times 10^{-37} \dfrac{\text{milliwatts}}{\text{atom(Roentgen/hr)}}$

Values of gamma heating calculated using these constants are shown in Figure 4.3 as the dashed curve. These final values are estimated to be accurate to ± 20 percent.

To account for self-shielding effects, one introduces a semiempirical relationship

$$A_0 = e^{-kd\left[\dfrac{(\frac{\rho}{A})(a_1 Z + a_2 Z^2 + a_3 Z^5)}{(\frac{\rho}{A})_{Pb}(a_1 Z + a_2 Z^2 + a_3 Z^5)_{Pb}}\right]} \qquad (4.16)$$

where d = Gaussian chord length = $\frac{4V}{s}$ where V is the volume and s is the surface area. From the data of Table 4.2, $k = 0.788\ \text{cm}^{-1}$ (d expressed in cm). The fit to the data is shown in Figure 4.4. Inserting numerical values into Equation (4.16) gives

$$A_0 = e^{-[0.060d\ (\frac{\rho}{A})(Z+5.00 \times 10^{-3} Z^2 + 3.42 \times 10^{-8} Z^5)]} \qquad (4.17)$$

where the density ρ is expressed in gm/cm^3 and the length d in cm.

The final semiempirical formula for gamma heating is thus

$$G = G_0 A_0 f_p \frac{\text{milliwatts}}{\text{atom(Roentgen/hr)}} \qquad (4.18)$$

$$= (1.13 \times 10^{-29} Z + 5.62 \times 10^{-32} Z^2 + 3.87 \times 10^{-37} Z^5)$$

$$\exp[-0.060 d(\frac{\rho}{A})(Z + 5.00 \times 10^{-3} Z^2 + 3.42 \times 10^{-8} Z^5)] f_p$$

where f_p = position factor to correct for changes in energy spectrum with location with respect to the gamma radiation source. For the research reactor used for the previous measurement.

f_p = 1.0 at 4 in. from face of core in water
= 1.4 at core face
= 1.5 inside core

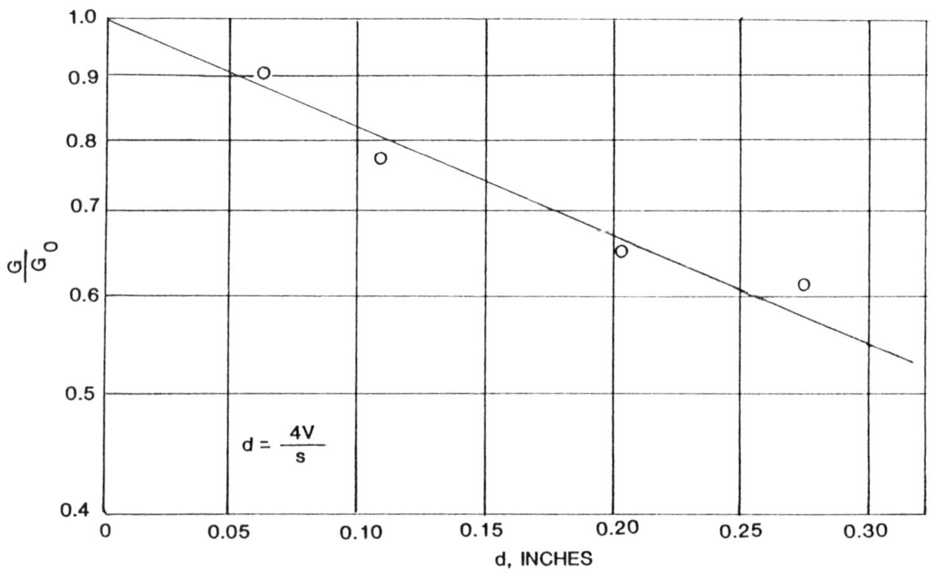

FIGURE 4.4 Gamma heating for lead samples at a reactor core face.

Illustration 4.3 Calculate the gamma heating in a thin 10 g piece of iron in a gamma field of 10^9 Roentgens/hr. Neglect self-shielding factor.

From Equation (4.18)

$$G_0 = 1.13 \times 10^{-29} Z + 5.62 \times 10^{-32} Z^2 + 3.87 \times 10^{-37} Z^5$$

For iron, Z = 26, so that

$$G_0 = 3.33 \times 10^{-28} \text{ milliwatts/atom per Roentgen/hr}$$

Hence

$$G_0 D = 3.33 \times 10^{-28} \times 10^9 = 3.33 \times 10^{-19} \text{ milliwatts/atom}$$

$$\text{Total number of atoms} = 10 \times \frac{6.02 \times 10^{23}}{55.85} = 1.08 \times 10^{23}$$

so that the gamma heating

$$= 1.08 \times 10^{23} \times 3.33 \times 10^{-19} = 3.6 \times 10^4 \text{ milliwatts} = 36 \text{ watts}$$

Illustration 4.4 Calculate the self-shielding factor in a 1/2-in. diameter long rod made of iron. From Equation (4.17) the self-shielding factor is

$$A_0 = \exp-[0.06d(\tfrac{\rho}{A})(Z + 5 \times 10^{-3} Z^2 + 3.42 \times 10^{-8} Z^5)]$$

where

$$d = \frac{4V}{s} = \frac{4(\pi \times \frac{D^2}{4} \times L)}{\pi DL + 2(\pi D^2/4)} = D \ (L >> D)$$

$$= 1.27 \text{ cm}$$

$$\rho = 7.86 \text{ g/cm}^3$$

$$A = 55.85 \text{ g/mole}$$

$$Z = 26$$

Introducing these values give

$$A_0 = 0.726$$

4.5 Relationship Between Gamma Flux and Dose Rate

The rad dose rate (or Gray dose rate) $\dot{D}$ is related to the flux of gamma ray photons as

$$\dot{D} = \Phi \frac{\mu_a}{\rho} E_\gamma \qquad (4.19)$$

where Φ is the photon flux [photons/(cm^2)(sec)], μ_a is the linear absorption coefficient, ρ is the physical density of the material and E_γ is the energy of the incident gamma radiation. The quantity (μ_a/ρ) is called the mass absorption coefficient, and is tabulated for various materials and gamma ray energies in Table 4.3 [Reference (4.7)]. In theory, if one knew the precise gamma ray energy to use in Equation (4.19), a single measurement of the dose rate (in air, for example) would determine the gamma flux, and hence, through the use of Table 4.3, would determine the dose rate (and hence gamma heating) for any other material exposed at that location. However, the principal problem with the use of Equation (4.19) is the assignment of a value to E_γ. For example, suppose an extensive material is exposed to cobalt-60 gamma radiation. Cobalt-60 emits two gamma rays, of energies 1.17 and 1.33 MeV (average 1.25 MeV). In the Compton interactions of these gammas with the material, there will be formed degraded gamma rays (see Figure 4.2), which will, in turn, undergo further energy-degrading reactions. Thus assignment of E_γ = 1.25 MeV for cobalt-60 radiations is incorrect. Further, as seen from Table 4.3, the mass absorption coefficient is a (weak) function of E_γ further requiring precise knowledge of E_γ. Because of these difficulties, one is led to use of empirical data, such as presented in Section 4.4 on gamma heating. However, Equation (4.19) is still useful for estimating purposes.

TABLE 4.3

The Mass Absorption Coefficient (μ_a/ρ) for Several Materials, in cm^2/g

Material	Gamma-Ray Energy, MeV																	
	0.1	0.15	0.2	0.3	0.4	0.5	0.6	0.8	1.0	1.25	1.50	2	3	4	5	6	8	10
H	.0411	.0487	.0531	.0575	.0589	.0591	.0590	.0575	.0557	.0533	.0509	.0467	.0401	.0354	.0318	.0291	.0252	.0255
Be	.0183	.0217	.0237	.0256	.0263	.0264	.0263	.0256	.0248	.0237	.0227	.0210	.0183	.0164	.0151	.0141	.0127	.0118
C	.0215	.0246	.0267	.0288	.0296	.0297	.0295	.0289	.0280	.0268	.0256	.0237	.0209	.0190	.0177	.0166	.0153	.0145
N	.0224	.0249	.0267	.0288	.0296	.0297	.0295	.0289	.0280	.0268	.0256	.0236	.0211	.0193	.0180	.0171	.0158	.0151
O	.0233	.0252	.0271	.0289	.0296	.0297	.0296	.0289	.0280	.0268	.0257	.0238	.0212	.0195	.0183	.0175	.0163	.0157
Na	.0289	.0258	.0266	.0279	.0283	.0284	.0284	.0276	.0268	.0257	.0246	.0229	.0207	.0194	.0185	.0179	.0171	.0168
Mg	.0335	.0276	.0278	.0290	.0294	.0293	.0292	.0285	.0276	.0265	.0254	.0237	.0215	.0203	.0194	.0188	.0182	.0180
Al	.0373	.0283	.0275	.0283	.0287	.0286	.0286	.0278	.0270	.0259	.0248	.0232	.0212	.0200	.0192	.0188	.0183	.0182
Si	.0435	.0300	.0286	.0291	.0293	.0290	.0290	.0282	.0274	.0263	.0252	.0236	.0217	.0206	.0198	.0194	.0190	.0189
P	.0501	.0315	.0292	.0289	.0290	.0290	.0287	.0280	.0271	.0260	.0250	.0234	.0216	.0206	.0200	.0197	.0194	.0195
S	.0601	.0351	.0310	.0301	.0301	.0300	.0298	.0288	.0279	.0268	.0258	.0242	.0224	.0215	.0209	.0206	.0206	.0206
Ar	.0729	.0368	.0302	.0278	.0274	.0272	.0270	.0260	.0252	.0242	.0233	.0220	.0206	.0199	.0195	.0195	.0194	.0197
K	.0909	.0433	.0340	.0304	.0298	.0295	.0291	.0282	.0272	.0261	.0251	.0237	.0222	.0217	.0214	.0212	.0215	.0219
Ca	.111	.0489	.0367	.0318	.0309	.0304	.0300	.0290	.0279	.0268	.0258	.0244	.0230	.0225	.0222	.0223	.0225	.0231
Fe	.225	.0810	.0489	.0340	.0307	.0294	.0287	.0274	.0261	.0250	.0242	.0231	.0224	.0224	.0227	.0231	.0239	.0250
Cu	.310	.107	.0594	.0368	.0316	.0296	.0286	.0271	.0260	.0247	.0237	.0229	.0223	.0227	.0231	.0237	.0248	.0261
Mo	.922	.294	.141	.0617	.0422	.0348	.0315	.0281	.0263	.0248	.0239	.0233	.0237	.0250	.0262	.0274	.0296	.0316
Sn	1.469	.471	.222	.0873	.0534	.0403	.0346	.0294	.0268	.0248	.0239	.0233	.0243	.0259	.0276	.0291	.0316	.0339
I	1.726	.557	.260	.100	.0589	.0433	.0366	.0303	.0274	.0252	.0241	.0236	.0247	.0265	.0283	.0299	.0327	.0353
W	4.112	1.356	.631	.230	.121	.0786	.0599	.0426	.0353	.0302	.0281	.0271	.0287	.0311	.0335	.0355	.0390	.0426
Pt	4.645	1.556	.719	.262	.138	.0892	.0666	.0465	.0375	.0315	.0293	.0280	.0296	.0320	.0343	.0365	.0400	.0438
Tl	5.057	1.717	.791	.285	.152	.0972	.0718	.0491	.0393	.0326	.0301	.0288	.0304	.0326	.0349	.0354	.0406	.0446
Pb	5.193	1.753	.821	.294	.156	.0994	.0738	.0505	.0402	.0332	.0306	.0293	.0305	.0330	.0352	.0373	.0412	.0450
U	9.63	2.337	1.096	.392	.208	.132	.0968	.0628	.0482	.0383	.0346	.0324	.0332	.0352	.0374	.0394	.0443	.0474
Air	.0233	.0251	.0268	.0288	.0296	.0297	.0296	.0289	.0280	.0268	.0256	.0238	.0211	.0194	.0181	.0172	.0160	.0153
NaI	1.466	.476	.224	.0889	.0542	.0410	.0354	.0299	.0273	.0253	.0242	.0235	.0241	.0254	.0268	.0281	.0303	.0325
H_2O	.0253	.0278	.0300	.0321	.0328	.0330	.0329	.0321	.0311	.0298	.0285	.0264	.0233	.0213	.0198	.0188	.0173	.0165
Concrete	.0416	.0300	.0289	.0294	.0297	.0296	.0295	.0287	.0278	.0272	.0256	.0239	.0216	.0203	.0194	.0188	.0180	.0177
Tissue	.0271	.0282	.0293	.0312	.0317	.0320	.0319	.0311	.0300	.0288	.0276	.0256	.0220	.0206	.0192	.0182	.0168	.0160

[Reproduced with permission from L.T. Templin, editor, Reactor Physics Constants, ANL-5800, 2nd ed., 1963]

4.6 Free Electron Generation

It is to be observed that the three major gamma ray interactions depicted in Figure 4.1, all lead to the formation of a free electron. This fact is important to several aspects of gamma radiation effects, which will be discussed in later sections of this text. It is also to be noted that the Compton-scattered gamma ray can cause further electron release. In general a number of electrons are freed for each incident gamma ray interaction. (Only the photoelectric effect annihilates gamma radiation.) Computer calculations based upon interaction probabilities were used to follow the fate of interacting gamma rays and their associated production of free electrons. Cross section data were obtained from various sources, primarily Davisson and Evans [Reference (4-8)]. The calculated average number of free electrons generated per primary gamma ray energies is presented in Table 4.4. It is seen from these calculations, that for typical 1-MeV reactor gamma rays, the number of freed electrons per interaction significantly exceeds unity.

TABLE 4.4

Number of High-Energy Electrons (>0.5 KeV) Produced in Various Materials by Gamma Rays of Various Initial Energies

$_Z X$ \ E_γ (MeV)	0.10	0.15	0.20	0.30	0.40	0.50	0.60	0.80	1.00	1.25	1.50	2.00	3.00	4.00	5.00
$_1H$	32.18	34.25	36.69	36.25	38.09	38.99	39.81	39.92	40.79	40.84	41.02	41.36	41.55	41.72	41.96
$_4Be$	13.87	15.99	17.17	18.47	19.26	19.76	20.12	20.61	20.99	21.30	21.56	21.87	22.31	22.60	22.80
$_6C$	9.68	11.70	12.87	14.18	14.93	15.44	15.80	16.31	16.67	16.98	17.23	17.57	18.00	18.28	18.49
$_7N$	8.01	9.98	11.13	12.44	13.18	13.68	14.05	14.56	14.92	15.24	15.48	15.83	16.25	16.53	16.74
$_8O$	6.77	8.66	9.79	11.09	11.84	12.34	12.71	13.22	13.57	13.89	14.13	14.48	14.91	15.18	15.39
$_{11}Na$	4.43	6.09	7.15	8.42	9.15	9.65	10.02	10.53	10.88	11.21	11.44	11.79	12.22	12.50	12.70
$_{12}Mg$	3.92	5.50	6.54	7.79	8.52	9.02	9.39	9.89	10.25	10.57	10.81	11.16	11.59	11.87	12.07
$_{13}Al$	3.51	5.00	6.01	7.25	7.98	8.48	8.84	9.35	9.70	10.02	10.26	10.62	11.04	11.32	11.53
$_{14}Si$	3.16	4.58	5.56	6.78	7.50	7.99	8.36	8.87	9.22	9.54	9.79	10.14	10.56	10.84	11.05
$_{15}P$	2.88	4.21	5.16	6.36	7.08	7.58	7.94	8.45	8.80	9.12	9.37	9.72	10.14	10.42	10.63
$_{16}S$	2.63	3.88	4.81	5.99	6.71	7.20	7.56	8.07	8.42	8.74	9.99	9.34	9.76	10.05	10.25
$_{18}Ar$	2.25	3.35	4.21	5.36	6.07	6.54	6.91	7.42	7.72	8.10	8.34	8.70	9.12	9.39	9.60
$_{19}K$	2.10	3.13	3.96	5.08	5.79	6.27	6.63	7.14	7.49	7.82	8.06	8.42	8.83	9.11	9.32
$_{20}Ca$	1.97	2.93	3.73	4.84	5.53	6.02	6.38	6.88	7.23	7.56	7.80	8.16	8.58	8.86	9.07
$_{26}Fe$	1.50	2.12	2.75	3.71	4.37	4.84	5.19	5.69	6.04	6.37	6.61	6.97	5.38	7.66	7.87
$_{29}Cu$	1.37	1.87	2.42	3.31	3.94	4.40	4.75	5.25	5.59	5.93	6.17	6.53	6.95	7.22	7.43
$_{42}Mo$	1.13	1.34	1.62	2.23	2.74	3.15	3.47	3.95	4.29	4.64	4.88	5.25	5.66	5.94	6.12
$_{50}Sn$	1.08	1.21	1.39	1.86	2.30	2.67	2.97	3.43	3.77	4.14	4.36	4.73	5.15	5.41	5.62
$_{53}I$	1.06	1.18	1.34	1.76	2.17	2.53	2.82	3.27	3.60	3.98	4.21	4.57	4.99	5.26	5.46
$_{74}W$	1.03	1.07	1.14	1.34	1.59	1.84	2.08	2.47	2.77	3.21	3.42	3.77	4.22	4.46	4.64
$_{78}Pt$	1.02	1.06	1.12	1.30	1.52	1.76	1.98	2.36	2.66	3.12	3.30	3.66	4.12	4.35	4.53
$_{81}Tl$	1.02	1.06	1.11	1.27	1.48	1.70	1.91	2.28	2.58	3.04	3.21	3.59	4.04	4.28	4.45
$_{82}Pb$	1.92	1.95	1.19	1.26	1.46	1.68	1.89	2.26	2.55	3.02	3.19	3.56	4.03	4.25	4.44
$_{92}U$	1.02	1.04	1.08	1.20	1.35	1.53	1.72	2.05	2.33	2.83	2.98	3.35	3.83	4.04	4.21

REFERENCES

(4-1) Compton, A.H. and Allison, S.K., X-rays in Theory and Experiment, D. Von Nostrand Company, Inc., 1935.

(4-2) Evans, R.D., The Atomic Nucleus, McGraw-Hill Book Co., Inc., 1955.

(4-3) Kaplan, I., Nuclear Physics, Addison-Wesley Publishing Co, Inc., 1956.

(4-4) Heitler, W., The Quantum Theory of Radiation, Oxford University Press, 1954.

(4-5) Goldstein, H., Fundamental Aspects of Reactor Shielding, Addison-Wesley Publishing Co., Inc., 1959.

(4-6) Anno, J.N., "Factors Influencing the Effects of Reactor Radiations on Electrical Components," Nuclear Applications, 2, pp. 371-378, October, 1966.

(4-7) Templin, L.T., editor, Reactor Physics Constants, ANL-5800, 2nd ed., 1963; based on G.W. Grodstein, National Bureau of Standards Circula 583, 1957.

(4-8) Davisson, C.M. and Evans, R.D., "Gamma Ray Absorption Coefficients," Rev. Mod. Phys. 24, No. 2, pp. 79-107, April, 1952.

PROBLEMS FOR CHAPTER 4

1. A piece of aluminum is exposed to 10-KeV gamma rays and a piece of lead is exposed to 1-MeV gamma rays. Which has the higher cross section, (a) for the photoelectric effect? (b) for the Compton effect?

2. A piece of copper receives a radiation dose rate of 10^6 rads/hr due to gamma radiation. What is the gamma heating in this piece of material (watts/gm)?

3. One kg of iron is exposed to a gamma dose rate of 10^8 Roentgens/hr at a location 4 in. from the face of a PWR core. Neglecting self-shielding, estimate the gamma heating (watts) in the iron. Assume a density of 7.9 gm/cm^3.

4. A cylinder of lead is one inch in diameter and two inches long. It is exposed to a gamma dose rate of 10^8 Roentgens/hr (as measured by an ionization chamber) inside a reactor core. What is the average rad dose rate received by the cylinder?

5. A hemisphere of iron, radius of one inch, is exposed to a gamma dose rate of 5×10^8 Roentgens/hr (as measured by an ionization chamber) inside a reactor core. What is the average rad dose rate received by this piece?

6. Assuming $E_\gamma = 1$ MeV, calculate the rad dose received by a thin piece of lead exposed to a gamma flux of 10^{13} photons/(cm^2)(sec).

5 TRANSIENT EFFECTS OF RADIATION ON ELECTRICAL COMPONENTS

During the past decades, a number of experiments on radiation effects that can influence the output of electrical components exposed to a radiation environment have been performed at various nuclear facilities. These effects are mostly instantaneous rather than due to long-term degradation. It was found that electrons are released from metal surfaces when bombarded by gamma radiation. The gammas also ionize residual gas in the experiment system and produce heating in the components (see Chapter 4). The resistance of insulators may be drastically altered in a radiation field, principally by photoconductivity. The effects discussed here are limited to those for which some useful engineering data exist.

This chapter presents transient effects, i.e., effects which appear only in the presence of radiation and then die out when the radiation source is removed. These transient effects are related to the dose rate rather than to the total dose. Permanent effects, e.g. physical property changes, are discussed in following chapters of this text.

5.1 Gamma-Induced Electron Emission from Surfaces

That X-rays and gamma rays induce electron emission

from surfaces is well known [References (5-1) through (5-4)]. Further, the emission contains both high-energy and low-energy electron groups [Reference (5-4)]. This section of the chapter presents information on these two types of electron emission.

5.1.1 Secondary electrons. Presumably the low-energy electrons include secondary electrons, released as the high-energy electrons emerge from the surface. It has been observed [Reference (5-2)] that these secondary electrons have an energy spectrum approximately Maxwellian up to about 40 eV. A good fit to the data has been obtained by the relations

$$
\begin{aligned}
N(E) &= k \left(\frac{E}{E_0}\right)^{1/2} \exp\left[-\frac{1}{2}\left(\frac{E}{E_0}\right)\right] && 0 \le E \le 41 \text{ eV} \\
&= \frac{k}{(E/E_0)^2} && 41 \text{ eV} \le E
\end{aligned}
\qquad (5.1)
$$

where

$N(E)dE$ = the number of electrons with energy E in the interval E to E + dE

E_0 = most probable electron energy, approximately 4 eV

k = normalizing constant

The $1/E^2$ tail on the distribution is presumably an uncollided secondary group resulting from the fact that when the primary electron energy is much greater than the secondary, the production of secondaries per unit path length is inversely proportional to the square of the secondary energy. While many of the secondaries degrade in energy through collisions to assume the Maxwellian distribution at low energy, a few escape the material virtually unmodified. The energy distribution of Equation (5.1) is shown graphically in Figure 5.1. It is seen that the majority of the electrons are contained in that portion of the spectrum assigned a Maxwellian distribution. A few electrons are present at higher energies; approximately three percent have energies greater than 100 eV.

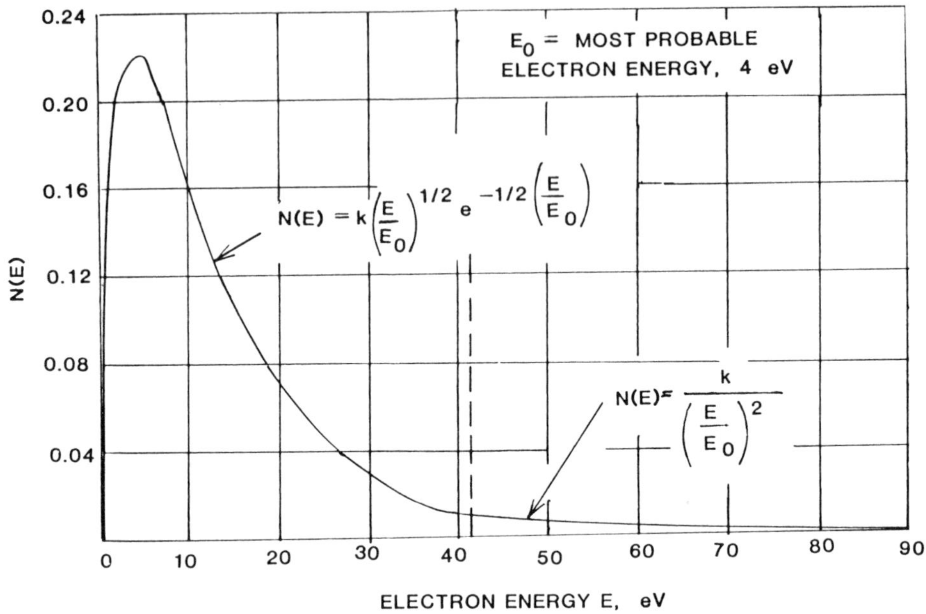

FIGURE 5.1 Energy spectrum of secondary electrons.

In a confirming experiment [Reference (5-5)] coaxial aluminum cylinders were exposed to reactor gamma radiation and the attendant electron emission measured. Under irradiation and with good vacuum conditions, the current to the central electrode was measured as voltage to the electrode was varied. The resulting current-voltage curve is shown in Figure 5.2 and compared with the predicted curve using Equation (5.1) showing excellent agreement. This experimental evidence is, of course, indirect indication of the electron-energy spectrum and detail is lost due to the geometrical transformation required. Certainly, nothing can be said as to the accuracy of the $1/E^2$ tail in describing the spectrum, since the electrons in this range contribute in a minor way to the current. However, the data fit is strong enough to support the prediction that the bulk of the electrons are described by a Maxwellian distribution with most probable energy near 4 eV.

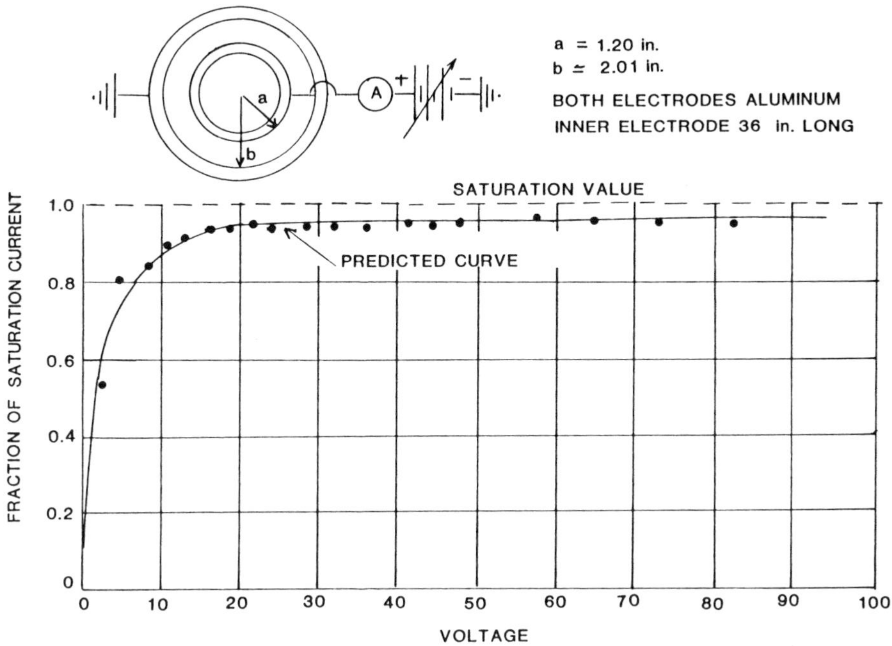

FIGURE 5.2 Comparison of predicted current-voltage curve with experiment results.

<u>5.1.2 High-energy electrons</u>. For reactor gamma radiations it has been found that for natural uranium metal, 66 percent of the electrons emitted had energies greater than 0.4 KeV [Reference (5-5)]. Similar findings by Wolcott [Reference (5-4)] showed that for ^{60}Co gamma irradiation of aluminum 70 percent of the electrons emitted had energies greater than 0.225 KeV. For gamma-induced electron

emission from a cylindrical UO_2-coated tube (3/4-in. diameter) surrounded by a suppressing squirrel-cage grid (2.40-in. diameter grid circle), it was found that the emitted reactor gamma-induced current I_γ varied with the magnitude of grid voltage V as [Reference (5-6)]

$$I_\gamma = k \ln \frac{V_0}{V} \tag{5.2}$$

where k is a constant and V_0 is of the order of 75 kilovolts. Thus it appeared that the maximum energy of the emitted electrons was of the order of 75 KeV. Table 5.1 presents calculations of the average energies $E_e(o)$ of the emerging electrons as a function of the incident gamma ray energy and the atomic number of the material. It is indeed seen that for ∿1-MeV gammas and intermediate Z materials, $E_e(o)$ is of the order of 75 KeV.

The interaction rate of gamma radiation with material is given as $\Sigma\Phi$ (primary gamma interactions)/(cm^3)(sec) where Σ is the macroscopic interaction cross section and Φ is the photon flux. Each primary gamma interaction produces a total of $\bar{\nu}$ free electrons [see Table 4.4], so that the volumetric source S of free electrons is

$$S = \Sigma\Phi\bar{\nu} \; \frac{\text{(electrons)}}{(\text{cm}^3)(\text{sec})} \tag{5.3}$$

For gamma radiation, Σ is simply the linear attenuation coefficient μ. Approximately one fourth of these electrons from within a mean range $\bar{r}$ of the surface escape, so that the flux of electrons crossing the surface is*

---

* One half of the electrons formed are directed outward toward the surface. Of these, only one half are within the proper cone of emission, which is bounded by the mean range $\bar{r}$. (See Appendix B) This expression neglects a (presumably) small contribution to the electron current from secondary electrons.

TABLE 5.1

Calculated Average Energies of Emerging Electrons for Various Materials and Various Initial Gamma Ray Energies

$_{Z}X$ \ E_γ (MeV)	$E_e(o)$ KeV										
	0.10	0.15	0.20	0.30	0.40	0.50	0.60	0.80	1.00	1.25	1.50
$_{4}Be$	4.91	6.03	7.53	10.9	14.4	17.8	21.1	27.5	33.5	40.7	47.4
$_{11}Na$	14.2	14.6	15.7	19.3	23.6	28.2	32.7	41.8	50.4	60.8	70.6
$_{12}Mg$	15.6	15.9	17.0	20.5	24.8	29.4	34.1	43.3	52.2	62.9	73.0
$_{13}Al$	16.9	17.2	18.2	21.7	26.0	30.7	35.4	44.9	54.0	64.9	75.3
$_{19}K$	24.1	25.0	25.8	28.8	33.1	37.9	43.0	53.4	63.5	75.8	87.5
$_{20}Ca$	25.1	26.2	27.0	30.0	34.3	39.1	44.3	54.7	65.0	77.4	89.3
$_{26}Fe$	29.5	33.0	34.3	37.2	41.3	46.2	51.4	62.3	73.3	86.6	99.5
$_{29}Cu$	30.7	35.8	37.7	40.1	44.8	49.6	54.9	66.0	77.2	90.9	104
$_{42}Mo$	31.1	42.5	48.8	54.9	59.4	64.3	69.6	81.2	93.1	108	123
$_{50}Sn$	28.6	42.7	52.0	61.7	67.4	72.7	78.2	90.0	102	118	133
$_{74}W$	13.6	32.2	47.6	69.5	82.8	92.0	99.4	113	126	143	160
$_{78}Pt$	9.88	29.1	45.2	68.8	83.8	93.9	102	116	130	147	164
$_{81}Tl$	6.85	26.5	43.1	68.0	84.1	95.1	104	118	132	150	167
$_{82}Pb$	5.79	25.6	42.3	67.6	84.1	95.5	104	119	133	151	168
$_{92}U$	--	14.9	33.2	62.2	82.6	97.0	108	125	140	159	176

$$\Phi_e = \frac{1}{4} S\bar{r} = \frac{\bar{r}\Sigma\Phi\bar{\nu}}{4} \quad \frac{\text{(electrons)}}{(\text{cm}^2)(\text{sec})} \tag{5.4}$$

As discussed in Chapter 4, the absorbed radiation dose rate in the material, $\dot{D}$, is related to the gamma flux as

$$\dot{D} = \frac{\Phi\mu_a E_\gamma}{\rho} \tag{5.5}$$

where μ_a is the linear absorption coefficient of the material, ρ its physical density, and E_γ is the initial energy of the gamma radiation. Combining Equations (5.4) and (5.5) gives the high-energy electron emission factor I_γ, amps/cm^2 per rad/hr as

$$I_\gamma = \frac{\Phi_e e}{\dot{D}} = \frac{\rho\bar{r}\bar{\nu}e}{4E_\gamma} \left(\frac{\mu}{\mu_a}\right) \tag{5.6}$$

To evaluate Equation (5.6), in addition to the calculations of $\bar{\nu}$ presented in Table 4.4, one needs data on the quantity $(\rho\bar{r})$ and (μ/μ_a). For a given initial energy of an electron E_e produced by the gamma interaction, the product $(\rho\bar{r})$ is a constant, given by [Reference (5-7)]

$$\rho\bar{r}\ \left(\frac{\text{gm}}{\text{cm}^2}\right) = \begin{cases} 0.412\ E_e^{(1.265-0.0954\ \ln E_e)} & E \leq 2.5\ \text{MeV} \\ 0.530\ E_e - 0.106 & E > 2.5\ \text{MeV} \end{cases} \tag{5.7}$$

for E_e expressed in MeV units. Data on the factor (μ/μ_a) were obtained from Reference (5-8) and are presented in Table 5.2.

Table 5.3 presents the results of calculations of I_γ for various incident gamma photon energies and for various Z materials. It is seen that, over the range of E_γ and Z displayed, the high-energy electron emission factors are of the same order of magnitude. For a given Z, I_γ increases with increasing E_γ, except for light materials ($Z < 13$) which exhibit a depression in I_γ near $E_\gamma = 0.3$ MeV. For a given E_γ, there is a maximum in I_γ in all

TABLE 5.2

$(\frac{\mu}{\mu_a})$ as a Function of Gamma Energy and Atomic Number Z

$_Z X$ \ E_γ (MeV)	0.10	0.15	0.20	0.30	0.40	0.50	0.60	0.80	1.00	1.25	1.50	2.00	3.00	4.00	5.00
$_1H$	7.19	5.41	4.56	3.69	3.20	2.92	2.72	2.43	2.26	2.12	2.02	1.88	1.72	1.64	1.58
$_4Be$	7.19	5.49	4.60	3.70	3.22	2.92	2.71	2.45	2.27	2.12	2.02	1.88	1.71	1.62	1.55
$_6C$	6.93	5.44	4.56	3.68	3.21	2.92	2.71	2.44	2.27	2.12	2.02	1.87	1.70	1.59	1.52
$_7N$	6.71	5.40	4.60	3.63	3.23	2.93	2.72	2.45	2.27	2.12	2.02	1.88	1.70	1.60	1.53
$_8O$	6.48	5.31	4.54	3.70	3.22	2.93	2.72	2.45	2.27	2.12	2.02	1.87	1.70	1.58	1.51
$_{11}Na$	5.22	5.03	4.43	3.65	3.22	2.93	2.71	2.44	2.26	2.12	2.01	1.86	1.68	1.56	1.48
$_{12}Mg$	4.77	4.89	4.38	3.65	3.21	2.93	2.72	2.45	2.27	2.11	2.01	1.86	1.67	1.55	1.47
$_{13}Al$	4.31	4.73	4.36	3.63	3.21	2.93	2.71	2.45	2.27	2.11	2.01	1.86	1.66	1.54	1.46
$_{14}Si$	3.95	4.63	4.37	3.67	3.25	2.99	2.77	2.50	2.32	2.16	2.05	1.89	1.69	1.56	1.49
$_{15}P$	3.47	4.35	4.18	3.60	3.20	2.92	2.72	2.45	2.28	2.12	2.01	1.86	1.66	1.53	1.45
$_{16}S$	3.13	4.11	4.09	3.58	3.18	2.91	2.70	2.45	2.28	2.12	2.01	1.85	1.66	1.53	1.45
$_{18}Ar$	2.58	3.67	3.87	3.51	3.16	2.90	2.70	2.45	2.27	2.12	2.01	1.85	1.64	1.51	1.43
$_{19}K$	2.36	3.44	3.73	3.48	3.14	2.88	2.70	2.44	2.27	2.12	2.01	1.84	1.64	1.50	1.42
$_{20}Ca$	2.14	3.23	3.59	3.42	3.12	2.88	2.69	2.44	2.27	2.11	2.00	1.84	1.63	1.50	1.42
$_{26}Fe$	1.52	2.25	2.82	3.11	2.99	2.81	2.65	2.42	2.27	2.12	2.00	1.83	1.61	1.47	1.37
$_{29}Cu$	1.37	1.92	2.47	2.93	2.89	2.77	2.62	2.41	2.25	2.10	2.00	1.82	1.60	1.45	1.36
$_{42}Mo$	1.11	1.32	1.59	2.10	2.36	2.44	2.41	2.30	2.18	2.05	1.95	1.77	1.54	1.39	1.31
$_{50}Sn$	1.07	1.19	1.36	1.75	2.04	2.19	2.24	2.20	2.11	2.02	1.92	1.75	1.51	1.37	1.28
$_{53}I$	1.02	1.06	1.12	1.27	1.43	1.59	1.68	1.79	1.81	1.99	1.91	1.73	1.50	1.36	1.28
$_{74}W$	1.02	1.06	1.12	1.27	1.43	1.59	1.68	1.79	1.81	1.80	1.75	1.61	1.41	1.29	1.22
$_{78}Pt$	1.02	1.08	1.10	1.23	1.38	1.51	1.60	1.72	1.75	1.75	1.70	1.58	1.39	1.28	1.21
$_{81}Tl$	1.02	1.04	1.09	1.21	1.34	1.47	1.55	1.67	1.71	1.72	1.68	1.56	1.38	1.27	1.21
$_{82}Pb$	1.01	1.04	1.09	1.21	1.33	1.45	1.54	1.65	1.70	1.71	1.67	1.55	1.38	1.27	1.21
$_{92}U$	1.10	1.03	1.06	1.10	1.24	1.33	1.40	1.51	1.57	1.60	1.58	1.49	1.34	1.25	1.19

[Reproduced with permission from L.T. Templin, editor, Reactor Physics Constants, ANL-5800, 2nd ed., 1963]

TABLE 5.3

Gamma-Induced High-Energy Electron Emission Factor,

I_γ: 10^{-16} amp/cm^2 per rad/hr

$_Z X$ \ E_γ (MeV)	0.10	0.15	0.20	0.30	0.40	0.50	0.60	0.80	1.00	1.25	1.50
$_4Be$	1.27	1.16	1.28	1.64	1.97	2.28	2.53	2.94	3.26	3.58	3.80
$_{11}Na$	2.68	2.49	2.25	2.21	2.38	2.57	2.74	3.11	3.38	3.65	3.84
$_{12}Mg$	2.62	2.61	2.38	2.30	2.43	2.63	2.80	3.14	3.38	3.63	3.86
$_{13}Al$	2.51	2.70	2.52	2.38	2.50	2.67	2.82	3.16	3.41	3.67	3.86
$_{19}K$	1.67	2.59	2.83	2.81	2.83	2.91	3.05	3.29	3.52	3.73	3.94
$_{20}Ca$	1.54	2.51	2.82	2.84	2.87	2.94	3.07	3.32	3.54	3.76	3.92
$_{26}Fe$	1.13	1.97	2.59	2.99	3.08	3.15	3.23	3.46	3.68	3.86	4.00
$_{29}Cu$	1.01	1.73	2.34	2.99	3.12	3.24	3.30	3.52	3.69	3.86	4.02
$_{42}Mo$	0.686	1.17	1.65	2.50	2.97	3.27	3.42	3.63	3.79	3.96	4.06
$_{50}Sn$	0.538	0.964	1.37	2.13	2.71	3.09	3.34	3.61	3.80	3.98	4.07
$_{74}W$	--	0.443	0.781	1.38	1.89	2.34	2.64	3.11	3.42	3.69	--
$_{78}Pt$	--	0.358	0.685	1.27	1.78	2.19	2.51	3.01	3.32	--	--
$_{81}Tl$	--	0.294	0.616	1.20	1.69	2.10	2.42	2.92	3.25	--	--
$_{82}Pb$	--	0.275	0.594	1.18	1.66	2.07	2.39	2.87	3.23	--	--
$_{92}U$	--	--	--	0.914	1.38	1.77	2.09	2.60	2.95	--	--

cases, being Z = 50 for E_γ = 1.25 MeV and dropping to Z = 13 for E_γ = 0.15 MeV. These calculated values are consistent with earlier observed values [References (5-2) and (5-3)].

Illustration 5.1 Calculate the high-energy gamma-induced current from 100 cm^2 of stainless steel receiving a gamma dose rate of 10^8 rads/hr of (initially) 1.25-MeV gamma rays.

From Table 5.3, the gamma-induced high-energy electron emission factor for iron for 1.25-Mev gammas is $I_\gamma = 3.86 \times 10^{-16}$ amps/cm^2 per rad/hr. Thus the induced current is

$$(3.86 \times 10^{-16})(100)(10^8) = 3.86 \times 10^{-6} \text{ amps.}$$

This illustration points out that gamma-induced currents from surfaces are small, generally in the range of microamperes.

5.1.3 Bulk emission factors. As with the gamma heating problem, discussed in Chapter 4, one of the problems in applying the calculations of I_γ presented in Table 5.3 is that one often does not know the rad dose rate, or the precise gamma ray energy but rather a gamma dose rate determined by an ionization chamber (air equivalent). For estimate purposes it is most convenient to rely upon experimental observations in some cases.

The following summarizes measurements of the electron yield from various materials bombarded by reactor gamma radiation in vacuum [Reference (5-3)]:

1. The electron emission factor for aluminum is 0.38×10^{-16} amps/cm^2 per Roentgen/hr of incident gamma radiation. This value is in good agreement with the observed emission from aluminum under bombardment of 150-KeV X-rays

[Reference (5-9)] and the general approximate value of 0.5 x 10^{-16} amps/cm^2 per Roentgen/hr derived for X-ray and gamma ray bombardment of an "electron metal" [Reference (5-2)]. From a thin coating of uranium, the electron emission was 6.1 x 10^{-16} amps/cm^2 per Roentgen/hr, more than an order of magnitude larger than that for aluminum. Copper and stainless steel values were in the range 1 to 2 x 10^{-16} amps/cm^2 per Roentgen/hr.

2. Both high-energy and low-energy electron groups are produced by gamma radiation. For a high-Z material, such as uranium, approximately twice as many high-energy (primary) electrons are produced as low-energy electrons.

3. The average energy of low-energy electron groups is 5 to 10 eV. The high-energy electron group apparently extends from approximately 1 to 100 KeV.

5.1.4 Radiation effects. This electron emission from surfaces exposed to gamma radiation can affect the operation of electrical apparatus in several ways:

1. The gamma-induced electron emission can cause a net DC current flow between electrodes, especially if they are composed of dissimilar metals.

2. Electrical insulators in vacuum can be "short-circuited" by gamma-induced electron ejection along the insulator surfaces.

3. Electrically isolated components in vacuum can acquire high potential as a result of the electron emission. For example, if an electrically isolated surface is emitting 10-KeV electrons under gamma bombardment, the voltage on the surface could increase to 10 KV.

4. The high-energy group of electrons emitted from surfaces by gammas can ionize residual gas atoms in the

vicinity.

Several design procedures will minimize this effect of gamma radiation:

1. Use of materials with low atomic number.

2. Minimize surface areas of electrodes and other sensitive components.

3. Eliminate unnecessary electrical isolation of components.

Sensitive components that fail to meet these criteria could be subject to faulty operation in a radiation environment.

5.2 Direct Ionization of Residual Gases by Gamma Radiation

A direct effect of the presence of a radiation environment is the ionization of the residual gases in the system. Gamma photons ionize gas atoms by removing an electron by a photoelectric or Compton collision. This ejected electron generally has sufficient energy itself to ionize gas atoms along its path. The degree of this secondary ionization, in fact, is so large that the primary ionization event is often negligible by comparison. The moving electron undergoes Coulomb interactions with the atomic electrons of the neighboring gas atoms, resulting in energy losses and in ionization of these neighboring atoms. Since, by definition, one Roentgen produces 2.08×10^9 ion pairs in 1 cm^3 of dry air at 0°C and 760 torr, the gamma-induced ionization to be expected, at pressure P (torr) is simply

$$I = (2.08 \times 10^9)\ P\ \dot{D}/(760) \text{ ion pairs}/(cm^3)(sec) \quad (5.8)$$

where $\dot{D}$ is the gamma dose rate in units of Roentgen/sec.

For dose rates encountered in or around the core of a reactor, ionization currents usually predominate over gamma-induced secondary-electron currents above 10^{-4} to 10^{-3} torr. Of course, whether or not either of these effects influences the current output from a component depends on the general magnitude of the out-of-pile current from the component.

To convert Equation (5.8) into units of electrical current experienced due to ionization, one needs to know the volume of gas V_g and the multiplication factor, M, of the ionization due to secondary ionization (avalanche effects) by the released electrons. Thus, in terms of current flow,

$$I = (2.08 \times 10^9)(\frac{P}{760})\ \dot{D}\ e\ V_g\ M \text{ amps} \qquad (5.9)$$

where e is the electronic charge (1.60×10^{-19} Coulombs/electron).

Illustration 5.2 Two parallel surfaces in a component exposed to radiation have 100 cm^3 of air "trapped" between them at atmospheric pressure. Because of gamma-induced electron emission, a voltage difference exists between the surfaces. The gamma dose rate is 10^6 Roentgens/hr as measured by an air ionization chamber at STP. What is the current flow between the surfaces due to ionization? (Neglect gas multiplication, i.e., avalanche effects.)

For proper use of Equation (5.9) the dose rate is to be expressed in Roentgen/sec. Hence $\dot{D} = \frac{10^6}{3600} = 278$ Roentgens/sec. Then, from Equation (5.9),

$$I = (2.08 \times 10^9)(1)(278)(1.6 \times 10^{-19})(100)(1)$$
$$= 9.25 \times 10^{-6} \text{ amps}$$

Note that, compared with Illustration 5.1, a higher current

results from ionization at a much lower dose rate than from direct gamma-induced electron emission from the surfaces. The simple design principle that this discussion and analyses suggest for designing radiation-resistant electrical components, is to minimize the trapped gas volume in the system.

5.3 Production of High-Energy Electrons by Thermal Neutron Reactions

Bombardment of a material by thermal neutrons in most cases produces (n,γ) reactions, forming daughter products which may undergo radioactive decay by emission of beta particles. The betas constitute another undesirable source of charged particles in apparatus sensitive to the flow of small currents and to ionization. The energy of the decay betas is usually of the order of several MeV. At saturation, the beta current per unit area is

$$(J_\beta)_{sat} = \frac{eR}{4} \sigma_a \Phi X_a \text{ amps/cm}^2 \qquad (5.10)$$

where

X_a = atomic concentration of material (mass number a) being bombarded by thermal neutrons, atom/cm^3
Φ = thermal neutron flux, n/(cm^2)(sec)
σ_a = microscopic thermal-neutron-activation cross section of parent material with mass number a, cm^2
R = range of betas in parent material, cm
e = electronic charge, Coulomb

The factor of 1/4 again accounts for the fact that approximately 1/4 of all betas formed escape from the surface, and is derived assuming linear paths of the betas in the material. (See Appendix B)

Illustration 5.3 Calculate the saturated thermal-neutron-induced current from an aluminum surface in a thermal neutron flux of $3 \times 10^{13} n/(cm^2)(sec)$. Assume moderator temperature to be 586°K. From Equation (5.10)

$$(J_\beta)_{sat} = \frac{eR}{4} \sigma_a \Phi X_a$$

where

$$e = 1.6 \times 10^{-19} \text{ Coulomb}$$

$$\Phi = 3 \times 10^{13} \text{ n/(cm}^2\text{)(sec)}$$

$$R = 0.5 \text{ cm}$$

$$\sigma_a = \sigma_{\substack{2200 \text{ m/sec} \\ 293°K}} \frac{\sqrt{\pi}}{2} \sqrt{\frac{293}{T}}$$

$$= (0.23 \times 10^{-24})\sqrt{\pi}/2 \sqrt{293/586}$$

$$= 1.45 \times 10^{-25} \text{ cm}^2$$

$$X_a = 6 \times 10^{22} \text{ atoms/cm}^3$$

Introducing these values gives

$$(J_\beta)_{sat} = 5.2 \times 10^{-9} \text{ amps/cm}^2$$

This is comparable to the gamma-induced electron emission at a dose rate of the order of 10^7 Roentgens/hr. Surfaces with dimensions of a few inches are thus capable of emitting beta currents of nearly a microampere when exposed to a thermal-neutron flux of 10^{13} n/(cm^2)(sec).

The consequences of these beta currents are similar to those discussed previously for gamma-induced electron emission. The principal difference is that the majority of the electrons from beta emission have high energy compared to the energy from the gamma-induced electrons, and hence the unattenuated betas can penetrate materials up to 1/4-in. thick. Note that, by Equation (5.10), the beta current is a strong function of material.

5.4 Effects of Gamma Radiation on Electrical Resistances and Insulators (Photoconductivity)

It is known that one effect of gamma radiation is to decrease the electrical resistivity of materials. This is particularly significant for those materials with high resistivity, such as glasses and a few of the plastics often used to insulate electrical leads. It is believed that the principal short-term electrical effect of the radiation is photo-induced conductance. This being assumed, gamma radiation and X-rays are the agents of primary concern. Data are sparse on gamma-induced conductance in materials and show considerable scatter. Several early surveys on this topic are available in the literature [References (5-9) through (5-11)] and there have been several rather exhaustive studies reported on special insulating materials [References (5-12) and (5-13)]. It is generally agreed that the photoconductivity under steady-state conditions is related to the dose rate by

$$\sigma_p = k\dot{D}^x \tag{5.11}$$

where σ_p is the photoconductivity in mhos per unit length, k is a proportionality constant, $\dot{D}$ is the dose rate, and x is a constant exponent for a given measurement. The values of x found by most experimenters range from about 0.5 to 1.0. Winslow and Alger [Reference (5-9)] found that x was less than 1.0, especially at "high" dose rate (15 kR/min).

A second facet of the photoconductivity is the time rate of change of conductivity under other than steady-state conditions, i.e., during the initial bombardment and also after the radiation is removed. Data obtained at the Federal Telecommunication Laboratories [References (5-10) and (5-14)] indicate that, at low gamma dose rates (10 Roentgens/hr), nearly a day is required for photoconductivity in some organic insulators to reach steady state after initiation of the radiation.

The detailed understanding of the relationship between the conductivity of a material and radiation dosage would require a much more extensive amount of research than has been performed in the past. However, a simple engineering model is proposed (and has been proposed previously) [Reference (5-15)] which suggests the general behavior of electrical resistances exposed to radiation.

5.4.1 Simplified analytical model. The electrical resistance R of a material is inversely related to its conductivity

$$R = C/\sigma_t \qquad (5.12)$$

where C is a constant dependent upon the geometry of the material. The total conductivity of a material exposed to radiation is the sum of the conductivity σ_d of the material in the absence of radiation and the photo-induced conductivity σ_p

$$\sigma_t = \sigma_d + \sigma_p \qquad (5.13)$$

Thus, Equation (5.12) can be written as

$$R = R_0 \left(\frac{\sigma_d}{\sigma_d + \sigma_p}\right) \qquad (5.14)$$

where R_0 is the electrical resistance of the material in the absence of radiation. The photo-induced conductivity can be written as

$$\sigma_p = \mu\ e\ n \qquad (5.15)$$

where μ is the electron mobility in the particular material, e is the electronic charge, and n is the density of photo-induced electrons.

To determine n, assume a simplified model expressible

by the differential equation

$$\frac{dn}{dt} = P - An^2 - BnN \qquad (5.16)$$

where P is the electron production term (proportional to the radiation dose rate $\dot{D}$), An^2 is the term for recombination of electrons and holes, and BnN is the term for trapping of electrons by a density of N traps of all kinds. In general, the constants A and B will depend on the particular material. It should be emphasized that Equation (5.16) is not meant to be a complete nor accurate description of the actual physical phenomena, but rather it is to be taken as a possibly useful engineering model with which to approach the problem.

If a constant radiation dosage is applied at time t = 0, Equation (5.16) has its solution

$$n(t) = \frac{2P}{BN} \frac{(1-e^{-\alpha BNt})}{(1-e^{-\alpha BNt}) + \alpha(1+e^{-\alpha BNt})} \qquad (5.17)$$

where

$$\alpha = \left(1 + \frac{4PA}{B^2N^2}\right)^{1/2} \qquad (5.18)$$

At equilibrium

$$n_\infty = \frac{2P}{BN(1+\alpha)} \qquad (5.19)$$

Consider two extreme cases:

a. Large dose rate (or small number of traps) such that $\frac{4PA}{B^2N^2} >> 1$. Then, Equation (5.19) becomes

$$n_\infty = \sqrt{\frac{P}{A}} \qquad (5.20)$$

i.e., the photo-induced conductivity is proportional to the square root of the dose rate $\dot{D}$.

b. Small dose rate (or large number of traps) such that $\frac{4PA}{B^2N^2} << 1$. Then, Equation (5.19) becomes

$$n_\infty = \frac{P}{BN} \tag{5.21}$$

i.e., the photo-induced conductivity is directly proportional to the dose rate $\dot{D}$. Thus, this simple model is consistent with the experimental observations leading to Equation (5.11)

$$\sigma_p = k\dot{D}^x \tag{5.11}$$

where $0.5 \leq x \leq 1.0$. Upon removal of the source of radiation, after reaching a photo-electron density of n_0, the density decays according to

$$n(t) = \frac{BN}{A} \left(\frac{n_0}{n_0 + \frac{BN}{A}}\right) \frac{1}{\left[e^{BNt} - \left(\frac{n_0}{n_0 + \frac{BN}{A}}\right)\right]} \tag{5.22}$$

5.4.2 Results of a simple experiment. The measurement of photoconduction in materials is not a simple task. Because of ionization and radiation-induced surface emission phenomena, it is difficult to separate true photoconduction effects. The gamma-induced changes in the resistance of a voltage cable and Stupakopf seal attached at the top of a vacuum chamber near the core of a nuclear reactor (gamma dose rate near 10^3 Roentgens/hr) was briefly studied as a function of gamma dose rate [Reference (5-3)]. To determine ionization and electron emission effects, the cable and seal were potted in Ceresine wax, the resistance of which was known from previous experience to be insensitive to gamma radiation at the dose rates encountered in the experiment. This wax was applied on both sides of the vacuum seal. The results are shown in Figure 5.3 in the form of current-voltage curves for various dose rates up to 2,000 Roentgens/hr. As seen from these results, the resistance

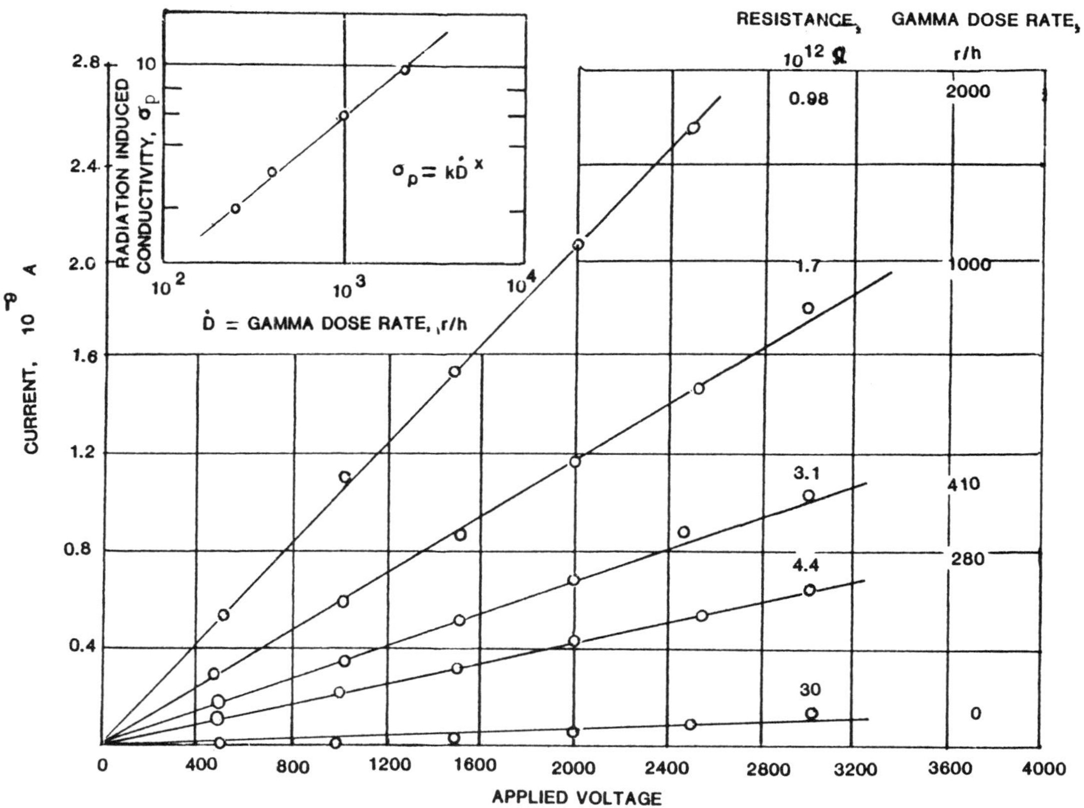

FIGURE 5.3 Current-voltage behavior of a cable-seal connection to a high vacuum facility for various gamma dose rates.

[Reproduced with permission from J.N. Anno, "Factors Influencing the Effects of Reactor Radiation on Electrical Components," Nuclear Applications, 2, pp. 371-378, October, 1966] of the cable and seal system decreased from 3×10^{13} ohms out-of-pile to 9.8×10^{11} ohms at 2,000 Roentgens/hr. The reciprocal of the resistance, with the out-of-pile contribution subtracted, should be proportional to the radiation-induced conductivity. The insert graph in Figure 5.3 shows a log-log plot of this conductivity, to within an unknown multiplier, versus the gamma dose rate. The graph shows that the data fit well to the simple relationship given in Equation (5.11) with x = 0.8. This value for x is within the range 0.5-1.0 suggested by the photoconductivity model and agrees with the conclusions of Winslow and Alger [Reference (5-9)]. Further, an exponent of 0.8 was found for electron irradiation of Mylar [Reference (5-16)]. These findings emphasize that even at relatively low dose rates, the resistance of electrical insulators can be rather drastically affected by gamma or X-ray radiation. The solution to the problem of photoconductivity is simply to use circuits with out-of-radiation resistances that are low. This implies a large value of σ_d so that the addition of a σ_p component will make only a small change in $\sigma_t = \sigma_d + \sigma_p$ and, hence, in the resistance [see Equation (5.14)].

5.5 Experimental Illustration of Gamma and Neutron Effects on Electrical Components

A simple experiment performed at a pool-type research reactor illustrates some of the principal effects of gamma and neutron radiation discussed in this chapter [Reference (5-3)].

For this experiment, a Pyrex ring (cylinder) placed in a 4-in. i.d. vacuum chamber served as a high-resistance insulator between a high-voltage brass lead (bare) and an aluminum plate at ground potential (Figure 5.4). The Pyrex ring was 1.5-in. long, 2.5-in. o.d., and had a wall thickness of approximately 3/32 in. Each end of the ring was silvered to provide uniform electrical contact. The long lead wire exaggerated the more common situation found in electrical components, but it served a useful purpose in emphasizing the effects of radiation under discussion. The vacuum chamber was placed near the reactor core (the core centerline was several inches below the bottom of the test chamber) and the current between the lead wire and ground was measured for various conditions as voltage was applied from batteries. The results are shown in Figure 5.5.

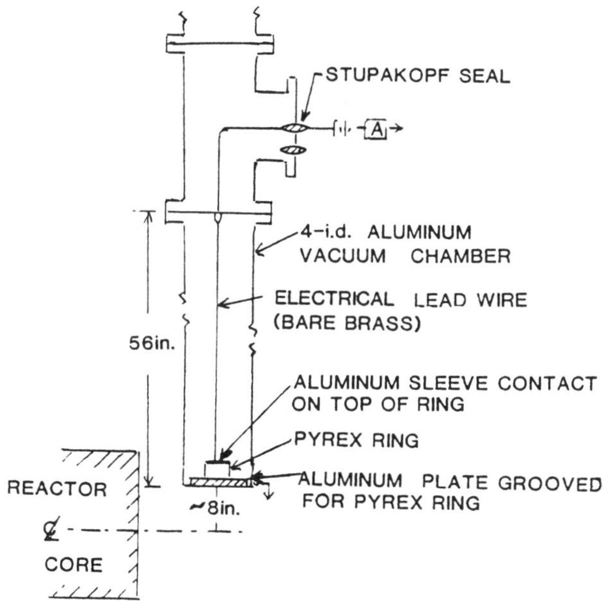

FIGURE 5.4 Experiment arrangement for vacuum irradiation of glass ring and lead line.

[Anno, Ref. (5-3)]

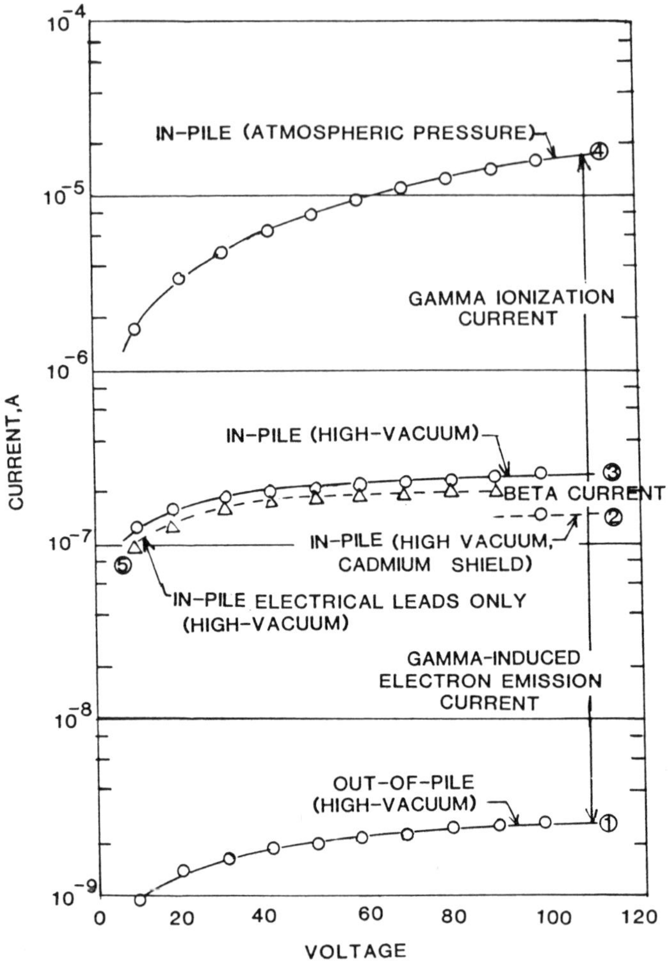

FIGURE 5.5 Current-voltage curves for "glass resistor" under various conditions.

[Anno, Ref. (5-3)]

Curve 1 shows the current-voltage curve for the experiment located far from the radiation source (core) with a moderately high vacuum (10^{-5} torr) in the test chamber. This curve includes all out-of-pile current leakage, so that the actual current flowing through the

Pyrex resistance was even smaller. Curve 3 shows the I-V curve at high vacuum with the experiment moved into a high radiation field adjacent to the core (gamma dose rate of the order of 10^7 Roentgens/hr. The radiation-induced increase in current might erroneously be presumed to be due to a deterioration of the resistance of the Pyrex ring as suggested in the previous section. However, curve 5 shows that nearly all of the change occurred in the electrical lead, itself; the Pyrex ring being completely removed from the chamber during this measurement. Curve 4 shows a further increase by two orders of magnitude in the current from the experiment when the pressure was increased to atmospheric pressure. In this case, the equipment was clearly functioning as an ion chamber and the measured I-V curve has no relation to the Pyrex resistance. Finally, for curve 2, the vacuum was restored but the data were obtained with a 20-mil-thick cadmium sleeve around the vacuum chamber to eliminate the thermal neutron component of the reactor radiation. The thermal neutron effect [presumably through beta production via decay of daughters from (n,γ) reactions] is seen to account for about half of the in-pile current at high vacuum.

From the comparisons given in Figure 5.5, it is seen that this particular experiment exhibits the gamma radiation effects discussed previously in this chapter and the neutron radiation effects, depending on the conditions of the irradiation. The fact that nearly all of the observed effects of radiation in this experiment are due to the electrical lead emphasizes the importance of eliminating extraneous effects in the design of electrical components for use in a radiation field, or in conducting radiation effects experiments with electrical apparatus.

5.6 Conclusions on the Transient Effects of Radiation on Electrical Components

Although there are other observations that can be made on the general effects of radiation on electrical components, the few presented here point out the care that must be taken in situations in which electrical components are exposed to radiation. Based upon these few observations presented here, the basic "rules of thumb" for design and use of electrical components in a radiation environment can be summarized as below:

a. Use low-Z materials to minimize the effects of gamma-induced electron current.

b. Minimize "free" surface area exposed to the gamma radiation.

c. Avoid electrically-isolated components (to prevent high voltage buildup).

d. Minimize "gas" volume in the electrical components to reduce the gamma ionization current.

e. Use low resistance circuits to minimize photoconductivity effects.

f. Use high current circuits to minimize photoconductivity effects.

g. Use materials with low neutron-absorption cross sections to minimize induced beta currents.

REFERENCES

(5-1) Taylor, L.S., "The Measurement of X- and Gamma Radiation Over a Wide Energy Range," Br. J. Radiol., 24, 75, 1951.

(5-2) Greening, J.R., "Contribution to the Theory of Ionization Chamber Measurements at Low Pressure," Br. J. Radiol., 27, 167, 1954.

(5-3) Anno, J.N., "Factors Influencing the Effects of Reactor Radiation on Electrical Components," Nuc. Applications, 2, 372, 1966.

(5-4) Wolcott, James D., and Anno, J.N., "Gamma-Induced Electron Emission from Conducting Surfaces," Nuc. Technol., 52, 139, January, 1981.

(5-5) Unpublished data by J.N. Anno, Battelle Memorial Institute, 1960.

(5-6) Unpublished data by S.L. Fawcett and J.N. Anno, Battelle Memorial Institute, 1960.

(5-7) Katz, L., and Penfold, A.S., "Range-Energy Relations for Electrons and the Determination of Beta-Ray End-Point Energies by Absorption," Rev. of Modern Phys., 24, No. 1, 32, 1952.

(5-8) Templin, L.T., Editor, Reactor Physics Constants, ANL-5800, 2nd Ed., 1963; based on G.W. Grodstein, National Bureau of Standards Circular 583, 1957.

(5-9) Winslow, J.W., and Alger, R.S., "Radiation-Induced Currents in Solid Organic Insulators," USNRDL-TR-326, U.S. Navy Radiological Defense Laboratory, 1959.

(5-10) Lamale, G.E., and Schall, P., Radiation Effects Information Center Memorandum on "Electrical Leakage in Insulators Exposed to a Nuclear Environment," January 15, 1958.

(5-11) Riley, W.C., Coppins, W.G., and Duckworth, W.H., Radiation Effects Information of Nuclear Center Technical Memorandum No. 9, "The Effect of Nuclear Radiation on Glass," November 30, 1958.

(5-12) Chrien, R.E., "Radiation-Induced Conductivity in Glass," Thesis, Rensselaer Polytechnic Institute, June, 1952.

(5-13) Culler, V.E., Rexford, H.E., and Barney, W.H., "Gamma Radiation Induced Conductivity in Solid Insulation," Paper No. CP 60-1085, AIEE Conference, 1960.

(5-14) "Ionization-Chamber Insulating Material," Federal Telecommunication Laboratories, Signal Corps Project No. 32-152B-0, July 15, 1951 through July 15, 1953.

(5-15) Kittel, Charles, Introduction to Solid State Physics, John Wiley & Sons, Inc., New York, 1956, Second Edition, pp. 513-514.

(5-16) Compton, D.M.J., Cheney, G.T., and Poll, R.A., "Radiation-Induced Conductivity in Plastic Films at High Dose Rates," J. Appl. Phys., 36, 2434, 1965.

PROBLEMS FOR CHAPTER 5

1. 100 cm^2 of tungsten (Z = 74) is exposed in vacuum to a gamma dose rate of 10^9 rads/hr. Estimate the current (amperes) due to high-energy electron emission from the tungsten.

2. A copper plate is exposed in vacuum to a cobalt-60 gamma dose rate of 10^8 Roentgen/hr.

   a. What is the current emission from the surface area of 100 cm^2 of this plate?

   b. Assuming that the energy of the electrons is very high (about 0.5 MeV) and that the plate is electrically insulated from ground with a 10^6-ohm resistance, what voltage will build up on the plate if it is in vacuum?

3. An electrical component has one liter of air at 2 atm pressure trapped between parallel surfaces with a voltage difference between them. The component is exposed to a gamma dose rate of 10^4 Roentgen/hr. What is the minimum expected current flow due to ionization?

4. A grounded aluminum plate is exposed in vacuum at room temperature to a gamma dose rate of 10^8 Roentgen/hr <u>and</u> a thermal neutron flux of 10^{13} n/(cm^2)(sec). What is the <u>total</u> current emission from the surface area of 100 cm^2 of this plate?

5. A glass electrical insulator, with "out-of-pile" resistance of 10^7 ohms, is exposed to a modest radiation field such that its resistance "in-pile" drops to 5 x 10^6 ohms. What fraction of the total conductivity is due to radiation-induced photoconductivity?

6 EFFECTS OF RADIATION ON ORGANIC MATERIALS

The use of plastics in electrical components is primarily as organic dielectric and insulating materials as applied to electrical circuitry. These materials are second only to semiconductor devices, such as integrated circuits, transistors, and diodes, in sensitivity to radiation as seen from the graph in Figure 2.1. Electrical and electronic circuit design must ultimately account for the sensitivity of components to radiation as one of the environmental conditions.

Degradation of plastics affects their electrical properties. Degradation of mechanical properties, however, is also a consideration to the extent that, in many applications, the mechanical failure of an insulator or dielectric will adversely affect its electrical characteristics. Radiation embrittlement of an insulator may lead to cracking and flaking, with a potential short-circuit hazard.

Organic insulating and dielectric materials experience both temporary and permanent changes in characteristics when subjected to a radiation environment. Data indicate that the temporary effects are generally rate sensitive with a saturation of the effect at the higher radiation levels. Permanent effects of radiation on organic insulating and dielectric materials are normally associated with a chemical change in the material. Most important among these chemical reactions that occur are molecular scission and crosslinkage. Scission is the breaking of bonds and usually causes a decrease in the average molecular weight. Crosslinking is the formation of new bonds between adjacent molecules and causes an increase in the average molecular weight. These chemical reactions or changes modify the physical properties of the material. A softening of the material decreases it in tensile strength and melting point, and a greater solubility could be the result of chain scission. Crosslinking leads to hardening, an increase in strength and melting point, a decrease in solubility, and an increase in density. Again, these permanent effects are changes in physical properties; changes in the dielectric dissipation factor and insulation resistance have also been recorded as permanent effects from radiation exposure, although these changes are often quite small.

A typical example for these competing effects in organic materials is the effect on the kinematic viscosity of SAE-30 motor oil when exposed to radiation. As seen in Figure 6.1 the viscosity decreases to a minimum and then increases [Reference (6-1)]. The breakpoint is at approximately 5 megarads, suggesting that scission is the dominant reaction before 5 megarad dose and crosslinking takes place primarily for dose larger than 10 megarads, for this particular dose rate of 7×10^4 rads/hr. Note that because of the competing effects, one would expect a dose rate dependency on this process, at least for a fluid. For a solid the effects tend to be "frozen in" and the radiation effects on physical properties are generally expressed in total dose, regardless of the dose rate. For further information on effects of radiation on lubricant viscosity, the reader is referred to Reference (6-2).

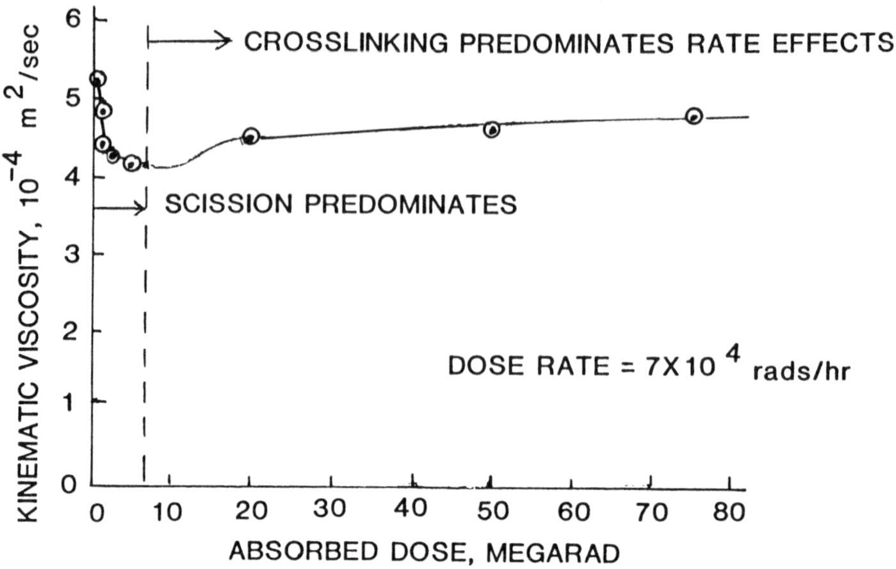

FIGURE 6.1 Viscosity of SAE-30 motor oil versus absorbed gamma dose.

[Reproduced with permission from Woon H. Chung and J.N. Anno, "Radiation-Induced Gas Evolution from Several Organic Liquids," Nuclear Technology, 52, 437-438, March, 1981]

6.1 Radiation Effects on Organic Insulating Materials

A comparison of the relative resistance of organic insulating materials to permanent effects [Reference (6-3)] is presented in Figure 6.2. The boundary between the open and crosshatched bars is identified with the damage threshold, while the boundary between the crosshatched and solid bars is identified with the 25% damage dose. These terms relate to changes in one or more physical properties, i.e., tensile strength, elongation, etc., with the damage threshold being the dose where the change is first detected. The 25% damage dose is that where a 25% change in the property occurs. It is seen that many of the insulating materials suffer damage to their physical properties at a dose of a few megarads.

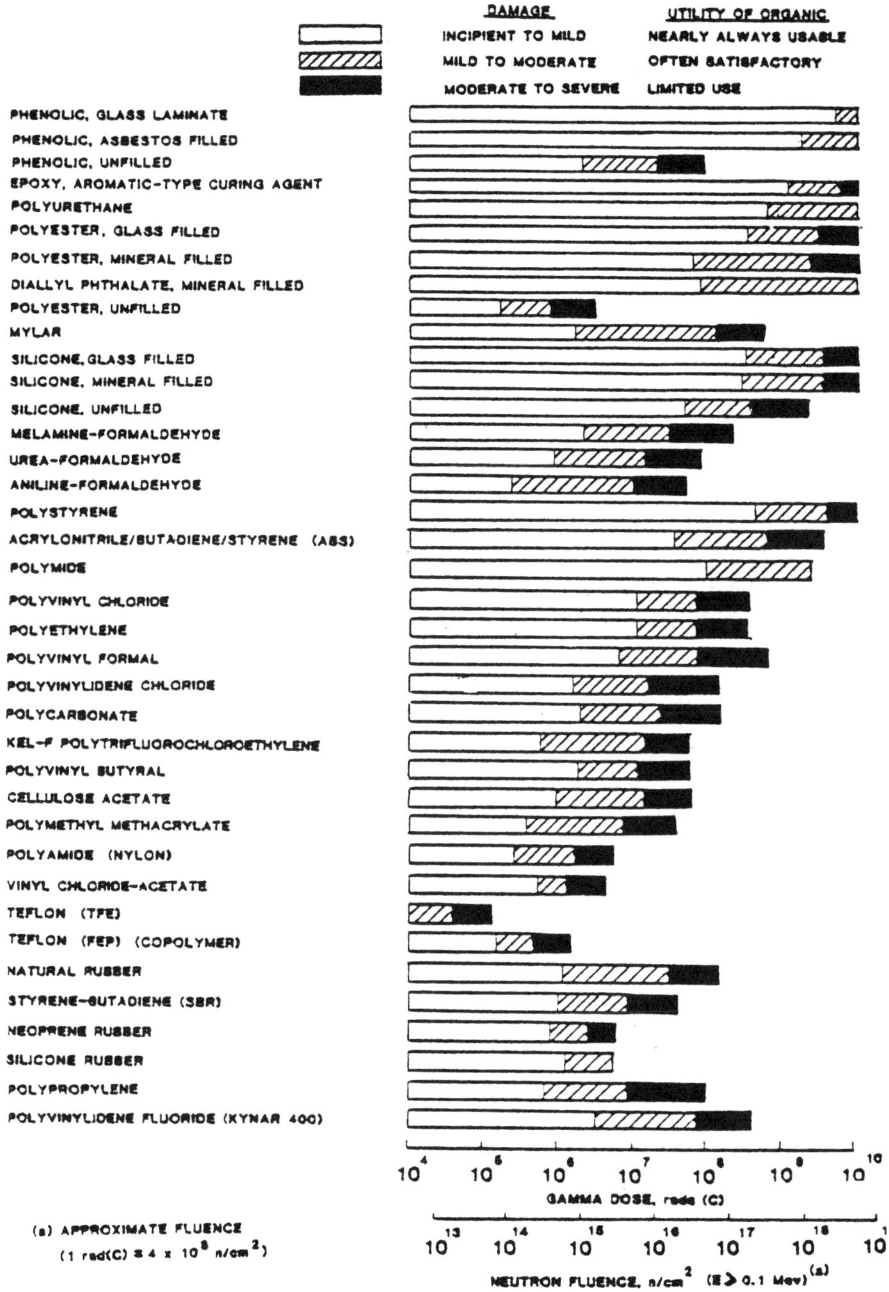

FIGURE 6.2 Relative radiation resistance of organic insulating materials based upon changes in physical properties.
[From REIC Report 21 and Addendum, Ref. (6-3)]

Illustration 6.1 A particular electrical connector comes with either Teflon or Polystyrene insulator. It is to be exposed to a gamma dose rate of 1 megarad/hr. How long will each material last until 25% damage occurs? From Figure 6.2 the dose of 25% damage for Teflon is 5×10^4 rads. Hence the time required is

$$\frac{5 \times 10^4}{10^6}$$

$$= 5 \times 10^{-2} \text{ hrs}$$

$$= 3 \text{ minutes}$$

For Polystyrene, the dose is 5×10^9 rads, hence it will last

$$\frac{5 \times 10^9}{10^6}$$

$$= 5000 \text{ hrs}$$

$$= 0.58 \text{ years}$$

6.2 Radiation-Induced Gas Evolution

Another reaction which may occur when an organic insulator or dielectric is irradiated is gas evolution. Gas evolution from solid organic polymers is less than that for liquids because of a greater recombination probability and a limited diffusion. It is unlikely, therefore, that the volume of evolved gas would be of serious concern, except possibly for a sealed enclosure filled with an organic fluid. Another problem with some evolved-gas species is that they are corrosive. This is true of the gases produced during the irradiation of halogenated hydrocarbons, such as polytetrafluoroethylene (Teflon) and polytrifluorochloroethylene (Kel-F). Although failure from other causes is likely to occur before the corrosion would become a problem, consideration of this problem is warranted for applications in sealed parts.

Radiolysis of organic liquids is well-documented

[Reference (6-4) and (6-5)]. However, quantitative measurements of the radiation-induced gas evolution are sparse. An experiment was conducted to find the radiation-induced gas evolution for several organic fluids (SAE-30 motor oil, brake fluid, transmission fluid, and vacuum pump oil) exposed to cobalt-60 gamma radiation [Reference (6-1)]. The gas evolution rate was determined by measuring the pressure rise in a closed system of known gas volume as a function of irradiation time. A typical experiment result is shown in Figure 6.3 for SAE-30 motor oil.

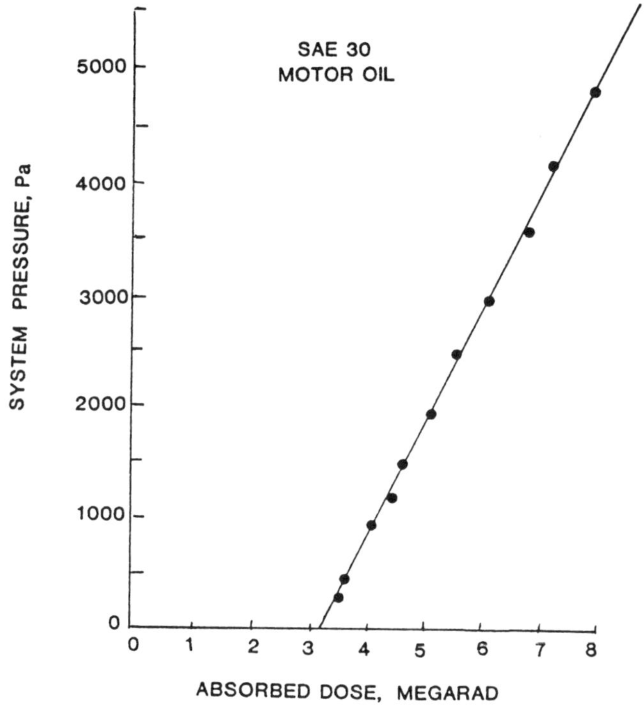

FIGURE 6.3 Pressure rise in a closed system versus absorbed dose.
[From Chung and Anno, Ref. (6-1)]

It was found in all cases that there was a definite threshold dose before significant pressure increase was detected and that the pressure rise was linear, indicating constant gas evolution rate. The results of the measurements are summarized in Table 6.1. All gas evolution rates were within a factor of five from each other, varying from 1.60 x 10^{-3} liter (gas)/liter (fluid) per megarad above threshold for the SAE-30 motor oil to 7.07 x 10^{-3} liter (gas)/liter (fluid) per megarad above threshold for the brake fluid. From dimensional analyses, the gas evolution was expressed in the semiempirical formula

$$E = 1.43 \times 10^{-12} \frac{\rho^{4/3} W^{2/3}}{\mu^2} + 1.58 \times 10^{-3} \qquad (6.1)$$

where E is the evolution rate in units of liter (gas)/liter (fluid) per megarad absorbed dose (above threshold), ρ is the fluid density (kg/m^3), W is the molecular weight, and μ is the absolute viscosity (Pa-sec). Equation (6.1) accurately describes the gas evolution for all fluids except the vacuum pump oil, in which case, the evolution is underestimated by a factor of 3.5.

No strong physical significance is given to the threshold dose, since it may indeed be related to the particular volume of fluid used (1.7 liters) and to the state of the fluid (amount of dissolved gases, etc.). However, it is interesting to note that the observed threshold dose Q_{th} for these particular fluids in this particular geometry is related to the molecular weight by the approximate formula

$$Q_{th}(\text{megarad}) = \frac{W-200}{100} \qquad (6.2)$$

Again, the vacuum pump oil is the exception to this formula. The reason(s) for the deviation of this oil from the behavior of the other fluids has not been determined, but may be due to the omission of other factors, such as solubility, surface tension, etc. from consideration. Detailed analyses of the constituents in the evolved gases were not performed. However, presumably, they consisted primarily of hydrogen, methane, and other carbon compounds.

TABLE 6.1

Summary of Experimental Determination of Gas Evolution Rates

Organic Liquid	Measured Molecular Weight	Measured Viscosity[a] Pa-sec	Density kg/m^3	Approximate Threshold Dose megarad	Gas Evolution Rate, 10^{-3} $\frac{\text{liter (gas)}}{\text{liter (fluid)}}$ per megarad absorbed dose above threshold
SAE-30 Motor Oil	524	0.340	884	3.05	1.60
Brake Fluid	294	0.011	1030	0.95	7.07
Transmission Fluid	395	0.049	868	2.03	1.83
Vacuum Pump	521	0.131	878	2.05	5.63

[a] Room temperature.

[From Chung and Anno, Ref. (6-1)]

6.3 Effects on Electrical Capacitors

Some data of the radiation effects on capacitors are given in Figures 6.4 through 6.7 [Reference (6-3)]. As seen in Figure 6.4, glass capacitors turn out to be much less sensitive to radiation than others.

6.4 Effects of Radiation on Epoxies

Table 6.2 gives the tolerance of radiation for selected epoxies [Reference (6-3)]. An interesting problem arises in applying epoxies as cable conduit filler in both a high temperature and high radiation environment. Under thermal aging, some epoxies shrink, breaking the potting seals between the electrical conductors and conduit walls [Reference (6-6)]. The idea was therefore conceived to counter the thermal aging shrinkage by adding a small portion of plastic filler to the epoxy which would swell under the influence of the radiation. Teflon, Nylon. Plexiglas and Polyethylene were chosen as candidate materials because of their relatively poor radiation resistance (see Figure 6.2). Approximately 5 volume percent of each plastic in the form of shredded small particles was added to specimens of epoxy casting resin, Stycast 2850FT. The specimens were then exposed to 100 megarads of cobalt-60 gamma radiation. The results are shown in Table 6.3 and indicate various degrees of radiation-induced swelling. This situation illustrates how radiation "damage" effects can be used to advantage under special conditions.

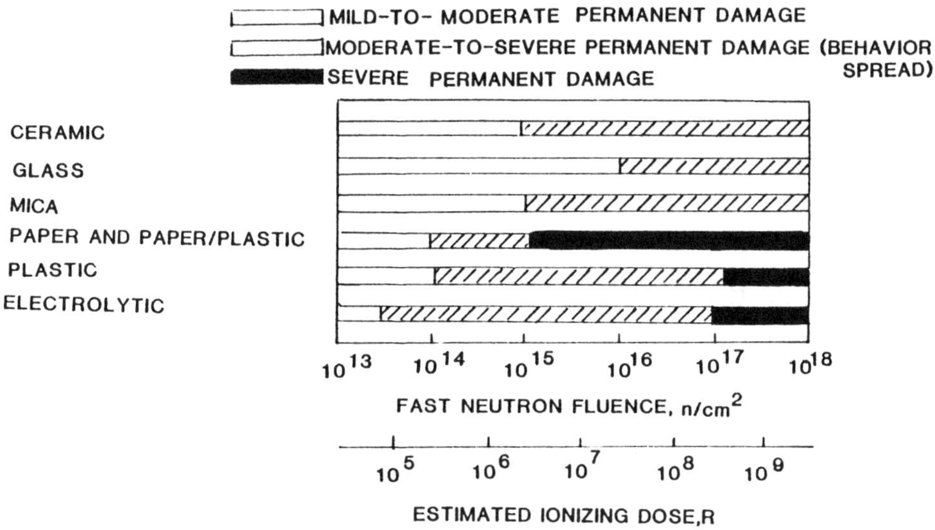

FIGURE 6.4 Relative radiation sensitivity of capacitors.
[Reproduced with permission from REIC Report No. 21 (and Addendum), "The Effect of Nuclear Radiation on Elastomeric and Plastic Components and Materials," Radiation Effects Information Center, Available from the National Technical Information Service, Springfield, VA 22161, as AD267890 and AD454056]

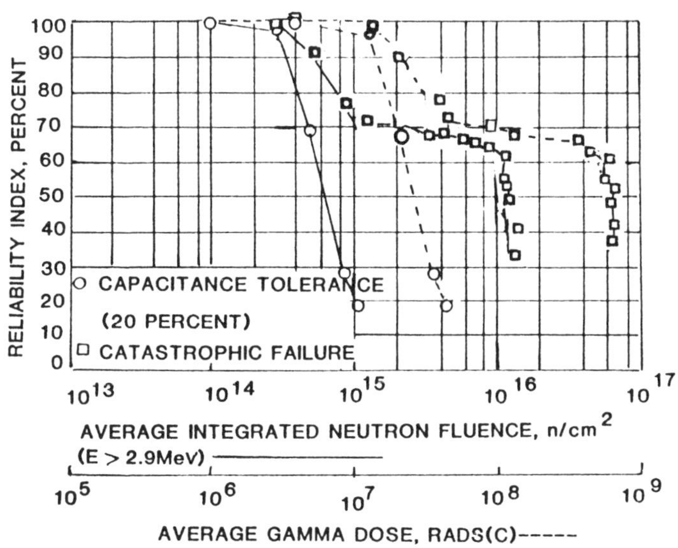

FIGURE 6.5 Reliability index for paper capacitors.
[From REIC Report No. 21 and Addendum, Ref. (6-3)]

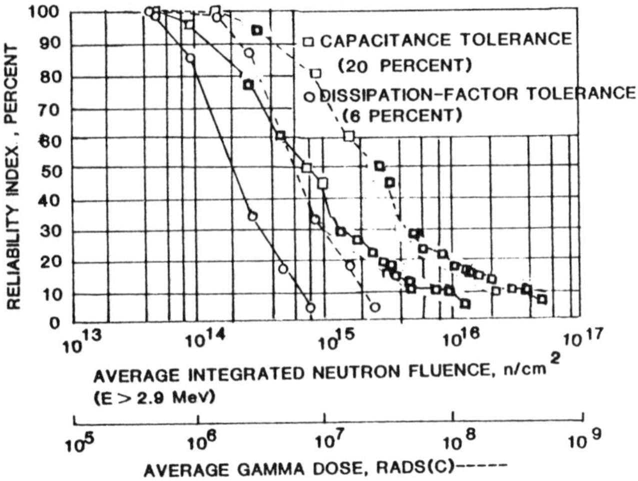

FIGURE 6.6 Reliability index for tantalum capacitors for a 95 percent confidence level, based on a sample size of 98 units (capacitance and dissipation factor).

[From REIC Report 21 and Addendum, Ref. (6-3)]

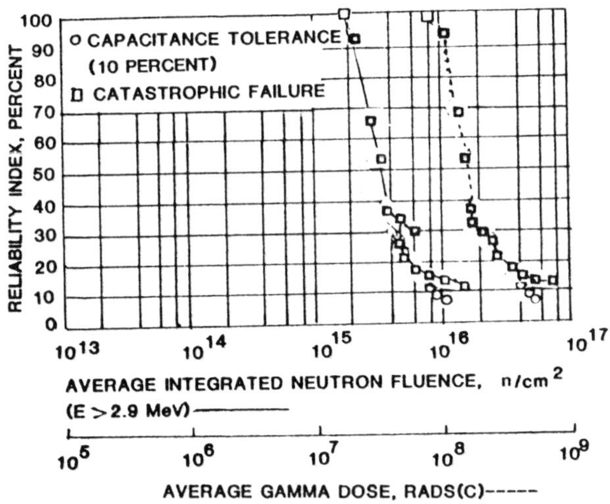

FIGURE 6.7 Reliability index for Mylar capacitors for a 95 percent confidence level, based on a sample size of 98 units (capacitance and dissipation factor).

[From REIC Report 21 and Addendum, Ref. (6-3)]

TABLE 6.2

Epoxies Exhibiting Satisfactory Radiation Tolerance at the Exposures Indicated

Epoxy Identification	Total Integrated Exposure[a]
Bisphenol A	8.8×10^7 rads (C) gamma 3.6×10^{13} n/cm^2 (E > 0.1 MeV)
Eccobond 182	1×10^8 rads (C) gamma 2×10^{13} n/cm^2 (E > 0.1 MeV)
Epocast 17B	8.8×10^7 rads (C) gamma 4.0×10^{13} n/cm^2 (E > 0.1 MeV)
Epon 828	4.4×10^6 rads (C) gamma 3.3×10^{15} n/cm^2 (E > 0.1 MeV)
Maraset 622-E	1×10^9 rads (C) gamma 1.1×10^{16} n/cm^2 (E > 0.5 MeV)
Novalak	8.8×10^7 rads (C) gamma 4.0×10^{13} n/cm^2 (E > 0.1 MeV)
Scotchcast 5	1×10^9 rads (C) gamma 1.1×10^{16} n/cm^2 (E > 0.5 MeV)
Scotchcast 212	1×10^9 rads (C) gamma 1.1×10^{16} n/cm^2 (E > 0.5 MeV)
Stycast 1095	1×10^8 rads (C) gamma 2×10^{13} n/cm^2 (E > 0.1 MeV)
Stycast 2651 MM	4.4×10^6 rads (C) gamma 3.3×10^{15} n/cm^2 (E > 0.1 MeV)
12-007	1.8×10^6 rads (C) gamma 1.5×10^{15} n/cm^2 (E > 0.1 MeV)
412-M	1×10^9 rads (C) gamma 1.1×10^{16} n/cm^2 (E > 0.5 MeV)
420-A	1×10^9 rads (C) gamma 1.1×10^{16} n/cm^2 (E > 0.5 MeV)
1126A/B	1.8×10^6 rads (C) gamma 1.5×10^{15} n/cm^2 (E > 0.1 MeV)
CF-8793	9.4×10^7 rads (C) gamma 3.8×10^{13} n/cm^2 (E > 0.1 MeV)
CF-8794	1.0×10^8 rads (C) gamma 4.0×10^{13} n/cm^2 (E > 0.1 MeV)
Unidentified (Mineral filled)	5.8×10^{16} e/cm^2 (E = 1.0 MeV)

(a) These exposures are not to be interpreted as indicating superiority in radiation tolerance of any material. They are the limits to which the materials have been subjected and are not damage thresholds.

[From REIC Report 21 and Addendum, Ref. (6-3)]

TABLE 6.3

Experimental Results of Resin Radiation-Induced Swelling with 5 V/o of Various Filler Materials

Filler	Percentage Volume Change[(a)]	Percentage Volume Change per V/o Filler per Megarad Absorbed Dose[(b)]
None	+ 0.7	--
Nylon	+ 2.4	0.0048
Polyethylene	+ 8.0	0.0160
Teflon	+ 10.4	0.0208
Polymethymethacrylate	+ 12.2	0.0244

(a) 100 ± 3 megarad dose of cobalt-60 gamma radiation.

(b) Assumes linearity of swelling with both V/o of filler and absorbed dose.

[Reproduced with permission from A.P. Newman, Cam-Lok Division, Empire Products, Inc., Cincinnati, OH, November, 1982; experiments by Research Dynamics Incorporated]

REFERENCES

(6-1) Chung, Woon H., and Anno, J.N., "Radiation-Induced Gas Evolution from Several Organic Liquids," Nuclear Technology, 52, 437-438, March, 1981.

(6-2) Bolt, R.O., and Carroll, J.G., Radiation Effects on Organic Materials, Academic Press, Inc., New York, 1963.

(6-3) REIC Report No. 21 (and Addendum), "The Effect of Nuclear Radiation on Elastomeric and Plastic Components and Materials," Radiation Effects Information Center, Battelle Memorial Institute, Columbus, Ohio, August 31, 1964, Available from the National Technical Information Service, Springfield, VA 22161, as AD267890 and AD454056.

(6-4) Sprinks, J.W.T., and Woods, R.J., An Introduction to Radiation Chemistry, 2nd ed., John Wiley and Sons, Inc., New York, 1976.

(6-5) Gaumann, T., and Horgne, J., Eds., Aspects of Hydrocarbon Radiolysis, Academic Press, Inc., London and New York, 1968.

(6-6) Private communication from A.P. Newman, Cam-Lok Division, Empire Products, Inc., Cincinnati, Ohio, November, 1982.

PROBLEMS FOR CHAPTER 6

1. Polyurethane is used in a mixed radiation environment of 10^{10} n/(cm^2)(sec) fast neutron flux and a gamma dose rate of 10^5 rads/hr. How long until the material reaches its radiation damage threshold?

2. A pump in the reactor cooling system loop is exposed to a gamma dose rate of 10^4 rads/hr. Approximately how long until the grease in the pump bearings will undergo radiation damage? (How often should the grease be changed?)

3. Two organic liquids have the same density and viscosity, but one has only half of the molecular weight of the other.

   a. From which would you expect the <u>more</u> outgassing?
   b. From which would you expect the outgassing <u>first</u>?

   Assume both liquids are under same irradiation condition.

4. A moving part is lubricated by SAE-30 oil. It is canned initially at 1 atm and room temperature. The lubricant volume in the can is 100 cm^3 and the gas volume is 100 cm^3. In operation the temperature of lubricant and trapped gas increase to 120 °C. It is then exposed to a gamma dose rate of 0.1 megarad/hr. How long will it take for the pressure in this can to reach 15 psig? (The volumetric expansion coefficient of oil is 0.00063 °C^{-1}.)

5. You must use either a glass or paper/plastic capacitor in a radiation environment. Based only upon considerations of radiation damage, which one would you choose?

7 RADIATION EFFECTS ON STEEL

Early in the development of the light-water fission reactor (LWR) technology, stainless steel was the predominant cladding material and is still a candidate for the cladding in the liquid metal fast breeder reactor (LMFBR). Further, stainless steel is a candidate for the first wall material for fusion reactors. Another important (to radiation effects studies) use of steel in LWRs is as the pressure vessel material. This chapter discusses selected aspects of the radiation effects on stainless steel and pressure vessel steel.

7.1 Radiation Effects on Strengths of Stainless Steel

Of a major concern are the effects of neutron irradiation on the strengths of stainless steel. Emphasis in this discussion is placed on the fracture and fatigue strengths. As background for this discussion, a brief presentation of general properties of stainless steels is first given.

7.1.1 General properties of stainless steel. Stainless steel is an iron-based alloy to which 10 percent or more of chromium is added. This material has good resistance to rusting, i.e., it is corrosion resistant. Different kinds of stainless steel are available in different series. The 300 series (austenitic) contains 18 w/o Cr, 8-12 w/o Ni, and up to 3 w/o Mo. Nickel is added to increase the corrosion resistance. The chromium content is very high to increase oxidation resistance. Examples of this series and frequently used stainless steels are SS-304 and SS-316. The composition of the 300 series steels is shown in Table 7.1 [Reference (7-1)]. The other principal stainless steel series is the 400 series (ferritic). An example is SS-440. Most steels of current interest in the nuclear industry belong to the 300 series (austenitic) so that the remainder of the discussion of this section will emphasize the properties of this series.

7.1.2 Mechanical properties. Mechanical properties of materials are those which a material has in response to mechanical forces. Mechanical properties include stress, strain, ductility, tensile strength, and elastic modulus. Because mechanical properties are so important they will be discussed in detail.

Consider first the stress-strain properties of steel. Stress is defined as force per unit area and is expressed in (psi, N/m^2). Strain is the deformation of a material and commonly expressed as a change in length per original length. Strain may be elastic or plastic. The elastic strain is reversible strain; it disappears after the stress is removed. A good example of elastic strain is a spring which is intended to change its shape under load, but returns to its original shape when the load is removed. The plastic strain is nonreversible.

A typical elastic stress-strain relationship is shown in Figure 7.1 [Reference (7-2)]. The elastic strain is linearly proportional to the amount of applied stress. This relation follows Hooke's law

TABLE 7.1

Composition of Austenitic Steel

AISI & ACI Standard Composition Ranges for Wrought & Cast Chromium-Nickel Stainless Steels
American Iron and Steel Institute Classification of Chromium-Nickel Stainless Steels

AISI Type	Composition, %								
	C max	Mn max	P max	S max	Si max	Cr	Ni	Mo	Other
201	0.15	5.50-7.50	.060	.030	1.00	16.00-18.00	3.50-5.50	--	N 0.25 max
202	0.15	7.50-10.00	.060	.030	1.00	17.00-19.00	4.00-6.00	--	N 0.25 max
301	0.15	2.00	.045	.030	1.00	16.00-18.00	6.00-8.00	--	--
302	0.15	2.00	.045	.030	1.00	17.00-19.00	8.00-10.00	--	--
302B	0.15	2.00	.045	.030	2.00-3.00	17.00-19.00	8.00-10.00	--	--
303	0.15	2.00	0.20	0.15 min	1.00	17.00-19.00	8.00-10.00	0.60 max	--
303Se	0.15	2.00	0.20	.06	1.00	17.00-19.00	8.00-10.00	--	Se 0.15 min
304	.08	2.00	.045	.030	1.00	18.00-20.00	8.00-12.00	--	--
304L	.03	2.00	.045	.030	1.00	18.00-20.00	8.00-12.00	--	--
305	0.12	2.00	.045	.030	1.00	17.00-19.00	10.00-13.00	--	--
308	.08	2.00	.045	.030	1.00	19.00-21.00	10.00-12.00	--	--
309	0.20	2.00	.045	.030	1.00	22.00-24.00	12.00-15.00	--	--
309S	.08	2.00	.045	.030	1.00	22.00-24.00	12.00-15.00	--	--
310	0.25	2.00	.045	.030	1.00	24.00-26.00	19.00-22.00	--	--
310S	.08	2.00	.045	.030	1.00	24.00-26.00	19.00-22.00	--	--
314	0.25	2.00	.045	.030	1.50-3.00	23.00-26.00	19.00-22.00	--	--
316	.08	2.00	.045	.030	1.00	16.00-18.00	10.00-14.00	2.00-3.00	--
316L	.03	2.00	.045	.030	1.00	16.00-18.00	10.00-14.00	2.00-3.00	--
317	.08	2.00	.045	.030	1.00	18.00-20.00	11.00-15.00	3.00-4.00	--
D319	.07	2.00	.045	.030	1.00	17.50-19.50	11.00-15.00	2.25-3.00	--
321	.08	2.00	.045	.030	1.00	17.00-19.00	9.00-12.00	--	Ti 5 x C min
347	.08	2.00	.045	.030	1.00	17.00-19.00	9.00-13.00	--	Cb-Ta 10 x C min
348	.08	2.00	.045	.030	1.00	17.00-19.00	9.00-13.00	--	Cb-Ta 10 x C min;Ta 0.10 max;Co 0.20 max

[Reproduced with permission from Brickner, K.G., "Stainless Steels for Room and Cryogenic Temperatures," Selection for Stainless Steels, American Society for Metals, 1968, p. 1-29]

$$\sigma = E\varepsilon \qquad (7.1)$$

where σ is the stress, E is the modulus of elasticity (Young's modulus), and ε is the elastic strain. The stress-strain relationship for austenitic steel is shown in Figure 7.2 [Reference (7-3)]. The ductile material has an elastic limit or proportional limit beyond which yield or plastic deformation occurs, i.e., the limit below which the stress-strain relation is linear. The ability of a material to resist plastic deformation is called the yield strength, which is a stress level related to the onset of irreversible plastic deformation. The tensile strength of a material is calculated by dividing the maximum load by the original cross sectional area. Because the cross sectional area of a ductile material may be reduced before it breaks, the breaking stress may be less than the true tensile strength. The true fracture stress is the load at fracture divided by the final cross sectional area.

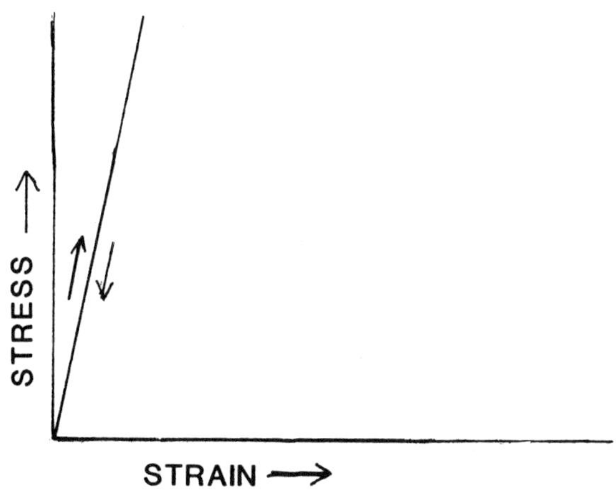

FIGURE 7.1 Elastic stress-strain relationship. Elastic strain is proportional to amount of stress.

[Van Vlack, L.H., Materials Science for Engineers, (c) 1970, Addison-Wesley Pub. Co. Inc., Reading, MA, page 8, Figure 1-2.1. Reprinted with permission]

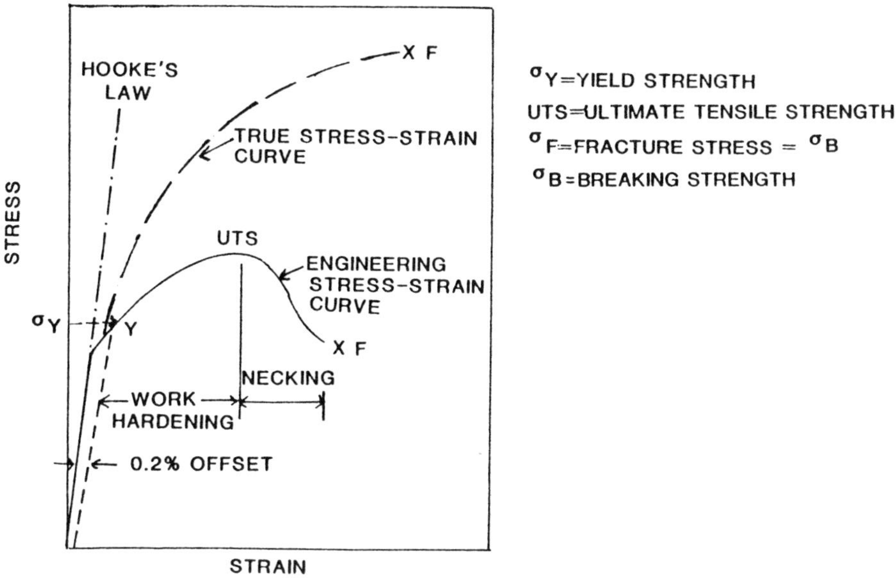

FIGURE 7.2 Stress-strain curve for austenitic steel.
[Reproduced with permission from D.R. Olander, Fundamental Aspect of Nuclear Reactor Fuel Elements, Energy Research and Development, Report TID-26711-P1, Ch. 18, 1976]

Another important feature is the ductility, which is the amount of plastic deformation at the breaking point and has the same units as the strain.

$$\text{Ductility} = \frac{A_0 - A_f}{A_0} \qquad (7.2)$$

where A_0 is the original area and A_f is the final area.

There are two types of stress-strain curves: the "true" and the "engineering" curves. The true stress in the material is based on actual area, while the engineering stress is based on original dimensions. Also the engineering stress is important for specifying the materials to be used in engineering products, while the true stress is important in the calculations of actual stress and deformation and to interpreting what is happening within the material itself. Before expressing the true and engineering stress and strain mathematically, let us imagine we have the specimen as shown in Figure 7.3 [Reference (7-4)],

where the notations in Figure 7.3 are as follows:

l_0 = original gauge length
A_0 = original cross sectional area
l_i = length at a certain stress (instantaneous)
A_i = cross sectional area at a certain stress (instantaneous)
P = load

The true strain depends on the instantaneous gauge length l_i

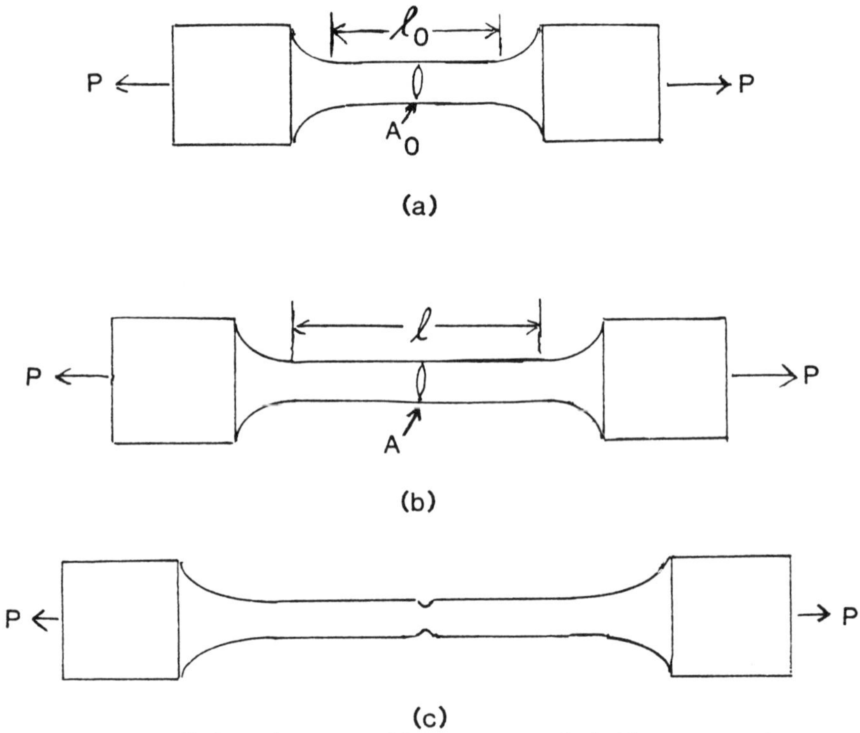

FIGURE 7.3 The tensile test. (a) Test specimen, (b) uniform elongation, (c) necking.

[Reproduced with permission from R. Hertzberg, Deformation and Fracture Mechanics of Engineering Materials, John Wiley and Sons, New York, Ch. 1, 7, and 12, 1976]

$$d\varepsilon = \frac{dl_i}{l_i} \quad (7.3)$$

Integrating,

$$\varepsilon_{true} = \int_0^{\varepsilon_{true}} d\varepsilon = \int_{l_0}^{l_f} \frac{dl_i}{l_i} = \ln \frac{l_f}{l_0} \quad (7.4)$$

where l_f is the final gauge length. Now the engineering strain, ε_{eng}, is defined as

$$\varepsilon_{eng} = \frac{l_f - l_0}{l_0} = \frac{\text{change in length}}{\text{original length}} \quad (7.5)$$

The engineering and true stress can be expressed as follows:

$$\sigma_{eng} = \text{engineering stress}$$

$$= \frac{\text{load}}{\text{original cross sectional area}} = \frac{P}{A_0} \quad (7.6)$$

$$\sigma_{true} = \text{true stress}$$

$$= \frac{\text{load}}{\text{instantaneous cross sectional area}} = \frac{P}{A_i} \quad (7.7)$$

The fundamental distinction concerning the definitions for true stress and strain is recognition of the interrelation between gauge length and diameter change. Since plastic deformation is a constant-volume process that

$$A_1 l_1 = A_2 l_2 \quad (7.8)$$

Consider a 10 cm long rod with an initial cross sectional area 1 cm^2 subjected to a force of 10^4 Newtons, the final area being reduced to 0.5 cm^2. In this case

$$\frac{A_1}{A_2} = \frac{l_2}{l_1} = 0.5 \text{ and } \sigma_{eng} = 10^8 \text{ N/m}^2,$$

whereas

$$\sigma_{true} = 2 \times 10^8 \text{ N/m}^2.$$

Thus the true stress is higher than the engineering stress.

The work hardening process, which is the resistance to further plastic deformation (as shown in Figure 7.2, the portion of true-stress curve from the onset of yielding to the maximum load) can be expressed by the Holloman relationship [Reference (7-5)].

$$\sigma_{true} = K\varepsilon^{n}_{true} \qquad (7.9)$$

where σ_{true}, ε_{true} = true stress and strain respectively

n = strain hardening coefficient (0.45-0.55 for stainless steel)

K = material constant, defined as the true stress at true strain of 1.0.

Another important factor in the stress-strain curve of Figure 7.2 is the necking phenomena. Necking generally begins at maximum load during the tensile deformation of a ductile material. At the necking instability the true plastic strain is numerically equal to the strain hardening coefficient, i.e.,

$$n = \varepsilon_{true} \qquad (7.10)$$

The necking instability will occur in uniaxial tension at a strain at which the slope of the true stress-strain curve equals the true stress at that strain. Mathematically this can be shown as follows. $P = A\sigma_{true}$ where P is the load, A is the area and σ_{true} = true stress. Differentiating the load expression, $dP = Ad\sigma_t + \sigma_t dA$. However, at necking dP = 0 so that

$$\frac{d\sigma_t}{\sigma_t} = \frac{-dA}{A}$$

Since plastic deformation is a constant volume process dV = 0 and d(AL) = AdL + LdA, so that

$$\frac{-dA}{A} = \frac{dL}{L}$$

But, the true strain is $d\varepsilon_{true} = \frac{dL}{L}$, and hence

$$\frac{d\sigma_t}{\sigma_t} = d\varepsilon_t .$$

The mechanical properties for the 300 series of stainless steels are shown in Table 7.2 [Reference (7-6)]. These properties are strong functions of temperature. The variation of several mechanical properties with temperature are sketched in Figure 7.4 [Reference (7-7)] and presented numerically in Tables 7.3 to 7.7 [Reference (7-8)]. It is clear from Figure 7.4 that an increase of temperature will cause a decrease in tensile strength and yield strength while the reduction of area (ductility) will increase with increasing temperature.

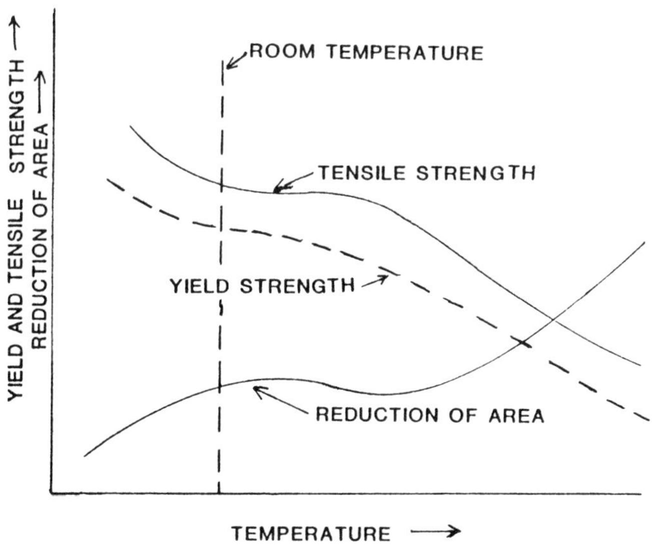

FIGURE 7.4 Variation of tensile properties of steel with temperature.

[Reproduced with permission from G.E. Dieter, Mechanical Metallurgy, McGraw-Hill Book Company, New York, Ch. 9, 1961]

TABLE 7.2

Tensile Properties of Annealed Chromium-Nickel Stainless Steels in Various Forms

AISI Type	Form	Tensile Strength, psi	Yield Strength (0.2% Offset), psi	Elongation (in 2 in.), %	Reduction of Area, %	Hardness	
						Brinell	Rockwell B
201	Sheet and Strip	95,000	38,000	40	--	--	90
202	Sheet	90,000	38,000	40	--	--	90
301	Sheet and Strip	75,000	30,000	40	--	--	--
	Plate	75,000	30,000	40	--	165	92
302	Sheet and Strip	75,000	30,000	40	--	--	92
	Plate	75,000	30,000	40	50	--	--
	Bars	75,000	30,000	40	50	150	--
302B	Sheet and Strip	75,000	30,000	40	--	--	85
	Plate	75,000	30,000	40	50	--	--
	Bars	75,000	30,000	40	50	--	--
303,303Se	Bars	85,000	36,000	50	50	262	--
304	Sheet and Strip	75,000	30,000	40	--	--	--
	Plate	75,000	30,000	40	--	--	--
	Bars	75,000	30,000	40	--	--	--
304L	Sheet and Strip	70,000	25,000	40	--	183	88
	Plate	70,000	25,000	40	--	--	--
305	Sheet and Strip	70,000	30,000	40	--	183	88
	Plate	70,000	25,000	40	--	183	88
308	Sheet and Strip	75,000	30,000	40	--	183	88
	Plate	75,000	30,000	40	--	--	--
	Bars	75,000	30,000	40	50	--	--
309,309S	Sheet and Strip	75,000	40,000	40	--	217	95
	Plate	75,000	40,000	40	--	217	95
	Bars	75,000	40,000	40	50	--	--

TABLE 7.2 (Continued)

AISI Type	Form	Tensile Strength, psi	Yield Strength (0.2% Offset), psi	Elongation (in 2 in.), %	Reduction of Area, %	Hardness	
						Brinell	Rockwell B
310,310S	Sheet and Strip	75,000	30,000	40	--	217	95
	Plate	75,000	30,000	40	65	217	95
	Bars	75,000	30,000	40	65	--	--
316	Sheet and Strip	75,000	30,000	40	--	217	95
	Plate	75,000	30,000	40	--	217	95
	Bars	75,000	30,000	40	50	--	--
316L	Sheet and Strip	70,000	25,000	40	--	217	95
	Plate	70,000	25,000	40	--	217	95
317	Sheet and Strip	75,000	30,000	--	--	217	95
	Plate	75,000	30,000	--	--	217	--
	Bars	75,000	40,000	50	--	--	--
321	Sheet and Strip	75,000	35,000	40	--	183	88
	Plate	75,000	30,000	40	--	183	88
	Bars	75,000	30,000	40	50	--	--
347,348	Sheet and Strip	75,000	30,000	40	--	183	88
	Plate	75,000	30,000	40	--	183	--
	Bars	75,000	30,000	40	50	--	--

[Reproduced with permission from the Steel Products Manual, Section 24, "Stainless Steel and Heat Resisting Steel," American Iron Steel Institute, December, 1974]

TABLE 7.3

Ratio of Yield Strength at Temperature to Yield Strength at Room Temperature for Wrought Austenitic Stainless Steels

Temperature		Grades					
°F	°C	304	304L	316	316L	321	347
75	24	1.00	1.00	1.00	1.00	1.00	1.00
100	38	0.96	0.97	0.97	0.97	0.97	0.98
200	93	0.83	0.85	0.86	0.85	0.85	0.92
300	149	0.75	0.77	0.78	0.76	0.76	0.85
400	204	0.69	0.70	0.70	0.70	0.69	0.80
500	260	0.65	0.65	0.66	0.64	0.64	0.75
600	316	0.61	0.62	0.63	0.61	0.61	0.72
700	371	0.59	0.60	0.60	0.58	0.59	0.69
800	427	0.56	0.58	0.59	0.55	0.57	0.68
900	482	0.54	0.56	0.58	0.53	0.57	0.67
1000	538	0.52	0.53	0.57	0.50	0.56	0.67
1100	593	0.49	0.50	0.55	0.47	0.55	0.66
1200	649	0.47	0.45	0.54	0.42	0.53	0.65
1300	704	0.44		0.51		0.50	0.61
1400	760	0.39		0.48		0.47	0.54
1500	816	0.31		0.43		0.40	0.43
1600	871	0.20					

[Reproduced with permission from J.H. Hoke, "Mechanical Properties of Stainless Steels at Elevated Temperature," in Handbook of Stainless Steel by D. Peckner and I.M. Bernstein, McGraw-Hill Book Company, New York, Ch. 21, 1977]

TABLE 7.4

Ratio of Tensile Strength at Temperature to Tensile Strength at Room Temperature for Wrought Austenitic Stainless Steels

Temperature		Grades					
°F	°C	304	304L	316	316L	321	347
75	24	1.00	1.00	1.00	1.00	1.00	1.00
100	38	0.97	0.97	0.97	0.97	0.97	0.97
200	93	0.86	0.86	0.91	0.88	0.89	0.87
300	149	0.80	0.79	0.89	0.83	0.84	0.80
400	204	0.78	0.76	0.87	0.81	0.83	0.75
500	260	0.77	0.75	0.87	0.80	0.83	0.73
600	316	0.77	0.74	0.88	0.80	0.83	0.72
700	371	0.77	0.73	0.87	0.80	0.83	0.71
800	427	0.76	0.72	0.86	0.79	0.83	0.71
900	482	0.74	0.70	0.83	0.76	0.83	0.71
1000	538	0.70	0.66	0.78	0.72	0.79	0.70
1100	593	0.63	0.60	0.71	0.66	0.71	0.67
1200	649	0.55	0.53	0.62	0.59	0.61	0.62
1300	704	0.46	0.45	0.52	0.50	0.49	0.54
1400	760	0.35	0.36	0.41	0.41	0.37	0.43
1500	816	0.25	0.26	0.30	0.31	0.25	0.28

[From Hoke, Ref. (7-8)]

TABLE 7.5

Effect of Temperature on Yield Strength, ksi (MN/m^2), of Wrought Austenitic Stainless Steels

Temperature		Grades					
°F	°C	304	304L	316	316L	321	347
75	24	36.9(254.4)	33.8(233.1)	38.0(262)	33.7(232.4)	29.7(204.8)	38.2(263.4)
100	38	35.4(244.1)	32.8(226.2)	36.9(254.4)	32.7(225.5)	28.8(198.6)	37.4(259.9)
200	93	30.6(211)	28.7(197.9)	32.7(225.5)	28.6(197.2)	25.2(173.8)	35.1(242)
300	149	27.7(191)	26.0(179.3)	29.7(204.8)	25.6(176.5)	22.6(155.8)	32.5(224.1)
400	204	25.5(175.8)	23.7(163.4)	26.6(183.4)	23.6(162.7)	20.5(141.3)	30.6(211)
500	260	24.0(165.5)	22.0(151.7)	25.1(173.1)	21.6(148.9)	19.0(131)	28.7(197.9)
600	316	22.5(155.1)	21.0(144.8)	23.9(164.8)	20.6(142)	18.1(124.8)	27.5(189.6)
700	371	21.8(150.3)	20.3(140)	22.8(157.2)	19.5(134.5)	17.5(120.7)	26.4(182)
800	427	20.7(142.7)	19.6(135.1)	22.4(154)	18.5(127.6)	16.9(116.5)	26.0(179.3)
900	482	19.9(137.2)	18.9(130.3)	22.0(151.7)	17.9(123.4)	16.9(116.5)	25.6(176.5)
1000	538	19.2(132.4)	17.9(123.4)	21.7(148.6)	16.9(116.5)	16.6(114.5)	25.6(176.5)
1100	593	18.1(124.8)	16.9(116.5)	20.9(144.1)	15.8(108.9)	16.3(112.4)	25.2(173.8)
1200	649	17.3(119.3)	15.2(104.8)	20.5(141.3)	14.2(97.9)	15.7(108.3)	24.8(171)
1300	704	16.2(111.7)		19.4(133.8)		14.8(102)	23.3(160.7)
1400	760	14.4(99.2)		18.2(125.5)		14.0(96.5)	20.6(142)
1500	816	11.4(78.6)		16.3(112.4)		13.7(94.5)	16.4(113.1)
1600	871	7.4(51)					

[From Hoke, Ref. (7-8)]

TABLE 7.6

Effect of Temperature on Ultimate Tensile Strength, ksi (MN/m^2), of Wrought Austenitic Stainless Steels

Temperature		Grades					
°F	°C	304	304L	316	316L	321	347
75	24	83.9(578.5)	79.2(546.1)	83.3(574.4)	78.9(544)	81.8(564)	87.0(600)
100	38	81.4(561.3)	76.8(529.5)	80.8(557.1)	76.5(527.5)	79.3(546.8)	84.4(581.9)
200	93	72.1(497.1)	68.1(469.5)	75.8(522.6)	69.4(478.5)	72.8(502)	75.7(522)
300	149	87.1(600.6)	62.6(431.6)	74.1(510.9)	65.5(451.6)	68.7(473.7)	69.6(480)
400	204	65.4(450.9)	60.2(415.1)	72.5(499.9)	63.9(440.6)	67.9(468.2)	65.3(450.2)
500	260	64.6(445.4)	59.4(409.6)	72.5(499.9)	63.1(435.1)	67.9(468.2)	63.5(437.8)
600	316	64.6(445.4)	58.6(404)	73.3(505.4)	63.1(435.1)	67.9(468.2)	62.7(432.3)
700	371	64.6(445.4)	57.8(398.5)	72.5(499.9)	63.1(435.1)	67.9(468.2)	61.8(426.1)
800	427	63.8(439.9)	57.0(393)	71.6(493.7)	62.3(429.6)	67.9(468.2)	61.8(426.1)
900	482	62.1(428.2)	55.4(382)	69.1(476.4)	60.0(413.7)	67.9(468.2)	61.8(426.1)
1000	538	58.7(404.7)	52.3(360.6)	65.0(448.2)	56.8(391.6)	64.6(445.4)	60.9(419.9)
1100	593	52.9(364.7)	47.5(327.5)	59.1(407.5)	52.1(359.2)	58.4(400.6)	58.3(402)
1200	649	46.1(317.9)	42.0(289.6)	51.6(355.8)	46.6(321.3)	50.0(344.8)	53.9(371.6)
1300	704	38.6(266.1)	35.6(245.5)	43.3(298.6)	39.5(272.4)	40.1(276.5)	47.0(324.1)
1400	760	29.4(202.7)	28.5(196.5)	34.2(235.8)	32.3(222.7)	30.3(208.9)	37.4(257.9)
1500	816	21.0(144.8)	20.6(142)	25.0(172.4)	24.5(168.9)	20.5(141.3)	24.4(168.2)

[From Hoke, Ref. (7-8)]

TABLE 7.7

Moduli of Elasticity at Various Temperatures for Several Austenitic Stainless Steels

Temperature			Grades					
°F	°C	Modulus*	302	304	310	316	321	347
200	93	E	27.9 (192.4)	27.9 (192.4)	28.2 (194.4)	28.1 (193.7)	28.0 (193.1)	28.2 (194.4)
		G	10.8 (74.5)	11.1 (76.5)	10.9 (75.2)	11.0 (75.8)	10.8 (74.5)	11.0 (75.8)
300	149	E	27.3 (188.2)	27.1 (186.9)	27.5 (189.6)	27.5 (189.6)	27.3 (188.2)	27.5 (189.6)
		G	10.4 (71.7)	10.8 (74.5)	10.6 (73.1)	10.6 (73.1)	10.6 (73.1)	10.7 (73.8)
400	204	E	26.7 (184.1)	26.6 (183.4)	26.8 (184.8)	26.9 (185.5)	26.5 (182.7)	26.8 (184.8)
		G	10.1 (69.6)	10.5 (72.4)	10.3 (71)	10.3 (71)	10.3 (71)	10.4 (71.7)
500	260	E	26.0 (179.3)	26.0 (179.3)	26.2 (180.6)	26.3 (181.3)	25.8 (177.9)	26.1 (180)
		G	9.8 (67.6)	10.2 (70.3)	10.0 (69)	10.0 (69)	9.9 (68.3)	10.1 (69.6)
600	316	E	25.4 (175.1)	25.6 (176.5)	25.5 (175.8)	25.6 (176.5)	25.3 (174.4)	25.4 (175.1)
		G	9.5 (65.5)	9.9 (68.3)	9.7 (66.9)	9.7 (66.9)	9.7 (66.9)	9.8 (67.6)
700	371	E	24.8 (171)	24.7 (170.3)	24.9 (171.7)	24.9 (171.7)	24.5 (168.9)	24.8 (171)
		G	9.3 (64.1)	9.7 (66.9)	9.4 (64.8)	9.4 (64.8)	9.4 (64.8)	9.5 (65.5)
800	427	E	24.2 (166.9)	24.1 (166.2)	24.2 (166.9)	24.2 (166.9)	23.8 (164.1)	24.1 (166.2)
		G	9.0 (62.1)	9.5 (65.5)	9.1 (62.7)	9.1 (62.7)	9.1 (62.7)	9.2 (63.4)
900	482	E	23.6 (162.7)	23.2 (160)	23.6 (162.7)	23.5 (162)	23.2 (160)	23.4 (161.3)
		G	8.8 (60.7)	9.2 (63.4)	8.8 (60.7)	8.8 (60.7)	8.8 (60.7)	8.9 (61.4)

TABLE 7.7 (Continued)

Temperature			Grades					
°F	°C	Modulus*	302	304	310	316	321	347
1000	538	E	23.0 (158.6)	22.5 (155.1)	23.0 (158.6)	22.8 (157.2)	22.5 (155.1)	22.8 (157.2)
		G	8.6 (59.3)	8.9 (61.4)	8.5 (58.6)	8.5 (58.6)	8.5 (58.6)	8.6 (59.3)
1110	593	E	22.3 (153.8)	21.8 (150.3)	22.4 (154.4)	22.2 (153.1)	21.9 (151)	22.0 (151.7)
		G	8.4 (57.9)	8.6 (59.3)	8.2 (56.5)	8.3 (57.2)	8.2 (56.5)	8.3 (57.3)
1200	649	E	21.8 (150.3)	21.1 (145.5)	21.8 (150.3)	21.5 (148.2)	21.2 (146.2)	21.4 (147.6)
		G	8.2 (56.5)	8.3 (57.2)	7.9 (54.5)	8.1 (55.8)	7.9 (54.5)	8.1 (55.8)
1300	704	E	21.2 (146.2)	20.4 (140.7)	21.2 (146.2)	20.8 (143.4)	20.4 (140.7)	20.7 (142.7)
		G	7.9 (54.5)	8.0 (55.2)	7.6 (52.4)	7.9 (54.5)	7.7 (53.1)	7.8 (53.9)
1400	760	E	20.6 (142)	19.4 (133.8)	20.5 (141.3)	20.0 (137.9)	19.7 (135.8)	20.0 (137.9)
		G	7.7 (53.1)	7.7 (53.1)	7.2 (49.6)	7.7 (53.1)	7.4 (51)	7.5 (51.7)
1500	816	E	20.0 (137.9)	18.1 (124.8)	19.0 (131)	19.1 (131.7)	19.1 (131.7)	19.4 (133.8)
		G	7.5 (51.7)	7.4 (51)	6.9 (47.6)	7.5 (51.7)	7.1 (49)	7.2 (49.6)

* E, modulus of elasticity in tension x 10^6 psi (GN/m^2). G, modulus of elasticity in shear x 10^6 psi (GN/m^3).

[From Hoke, Ref. (7-8)]

Next consider fatigue and fracture properties of the stainless steel. The term fatigue refers to the fracture arising from cyclic stresses, which usually are alternating between tension and compression. The term "fracture" can be defined as failure through applied stresses. The brittle fracture is the fracture which involves minimum energy absorption and plastic deformation. The mode of brittle fracture may be a "cleavage" along crystal planes, or it may be "intergranular" along grain boundaries. Another kind of fracture which is called "ductile fracture" involves considerable plastic deformation and possibility of internal rupture from defects before the actual fracture occurs. The mechanism of failure, particularly at high temperature, is called creep and stress rupture. The energy required (absorbed) before and during the fracture process is called "toughness" [Reference (7-4)], and is measured by the area under the stress-strain curve as illustrated in Figure 7.5. Mathematically,

$$\text{toughness energy} = \int_0^{\varepsilon_f} \sigma d\varepsilon \tag{7.11}$$

where ε_f is the final engineering strain and σ is the engineering stress.

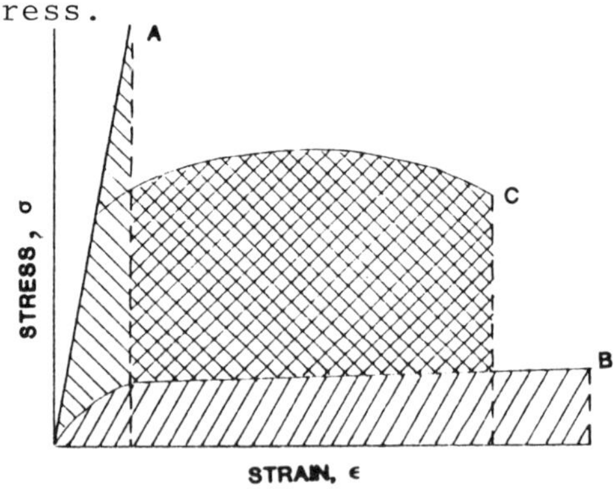

FIGURE 7.5 Stress-strain curves for strong material with little plastic flow capacity A, low strength but high ductility material B, and material with optimum combination of strength and ductility for maximum toughness C.
[From Hertzberg, Ref. (7-4)]

The S-N curve designates stress versus the number of cycles to failure. Figure 7.6 is the S-N curve for SS-304 [Reference (7-10)]. It is clear from the curve that the cyclic life increases with decreasing stress level and, further, below a certain stress level the material seems to possess "infinite life," i.e., fatigue failure will not occur for stress less than 110,000 psi [Reference (7-4)]. The presence of a notch will reduce the fatigue life as shown in Figure 7.6. The stress below which failure will not occur no matter what number of cycles is applied is called the endurance limit. The strain versus cycles to failure for SS-306 is shown in Figure 7.7 [Reference (7-10)]. In the region of lifetime from 10^3 to 10^6 cycles, the behavior of the S-N curve relating logarithm of stress amplitude to the logarithm of the number of cycles is nearly a straight line and can be described as follows

$$\sigma_a^m N = \text{constant} \tag{7.12}$$

where

σ_a = alternating stress

m = constant (8 to 15)

N = number of cycles to failure

<u>7.1.3 Radiation effects on the first wall stainless steel in a fusion reactor</u>. Generally the effect of neutron radiation on stainless steel as a first wall in a fusion power plant can be summarized in Table 7.8 [Reference (7-11)].

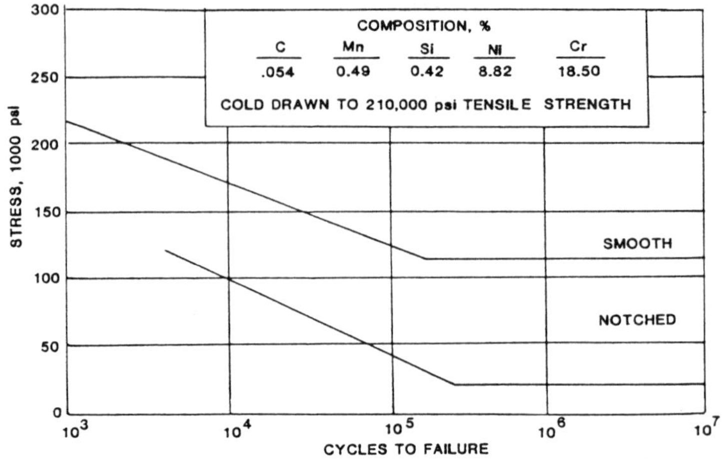

FIGURE 7.6 S-N curves obtained in vibrating beam tests notched and unnotched specimens of Type 304 stainless steel.

[Taken from: R. Hertzberg, Deformation and Fracture Mechanics of Engineering Materials, John Wiley and Sons, New York, 1976 with permission from R.E. Peterson, Edgar Marbury Lecture, ASTM, 1962; p.1]

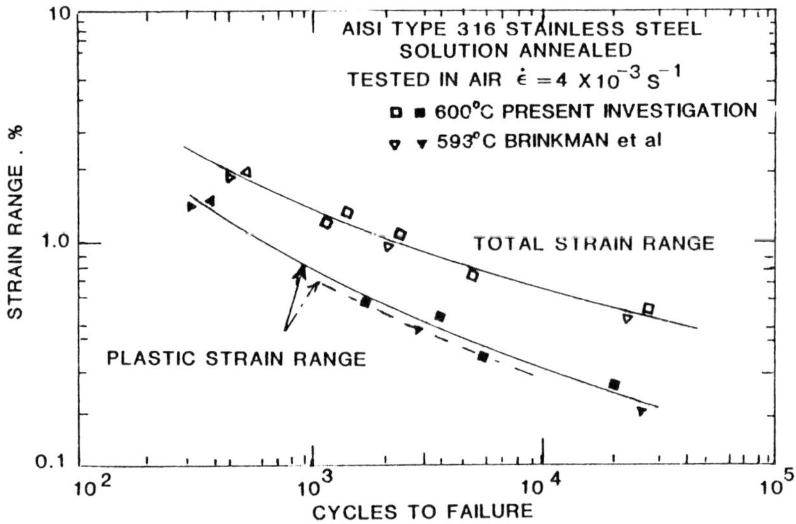

FIGURE 7.7 Total and plastic strain range versus cycles to failure for Type 316 stainless steel tested in air.

[Reproduced with permission from D.L. Smith, et al, "In Symposium on Structural Materials for Service at Elevated Temperatures in Nuclear Power Generation," ASME Winter Meeting, Nov.30-Dec.3, 1975, Houston, Texas, pp. 290-301]

TABLE 7.8

Schematic of the Response of Materials to Neutron Damage

Primary Response	Secondary Response
	1. Sputtering
A. Displacement	2. Swelling (voids, bubbles, growth)
	3. Resistance to change or effects on superconductivity
	4. Yield strength increase
B. Transmutations	5. Ductility loss
	6. Creep rate change
	7. Compositional change

[Reproduced with permission from H. Knoepfel, Eric-Trapani (Sicily), "Tokamak Reactors for Breakeven," Pergamon Press, 9/21-10/1-76, pp.449-84]

When energetic neutrons strike on solid materials, there are two responses: the primary and the secondary response. Generally, radiation will increase the yield strength and reduce the ductility as shown in Figure 7.8 [Reference (7-3)]. For Type 316 stainless steel the variation of properties with neutron fluence is shown in Figures 7.9 through 7.13, with further data presented in Table 7.9 [Reference (7-12)]. Fish and Holmes have found that the flow stress in annealed Type 316 stainless steel increases with increasing fast neutron fluence. The strengthening caused by fast neutrons decreases continuously from 480° to 700°C. No hardening occurs at or above 760°C. Reduction in area of Type 316 stainless steel decreases continuously as the fluence varies between 10^{22} n/cm^2 to 7 x 10^{22} n/cm^2(E_n>0.1 MeV) at temperature 540°C. The most drastic reductions in area occur above 650°C. In the region 430-480°C the variations of reduction in area with fast neutron fluence are less drastic.

Another serious radiation effect on the first wall material in a fusion reactor is that due to helium bombardment [Reference (7-13)] from the D-T reaction

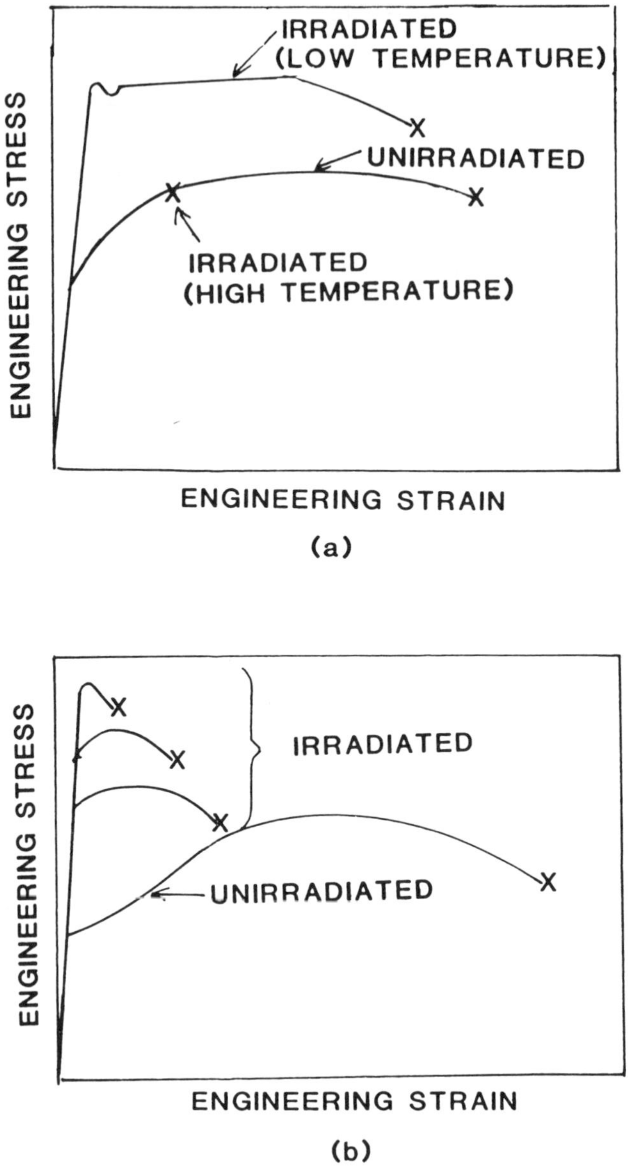

FIGURE 7.8 Effect of fast-neutron irradiation on the tensile properties of reactor steels. (a) Face-centered cubic structure. (b) Body-centered cubic structure.
[From Olander, Ref. (7-3)]

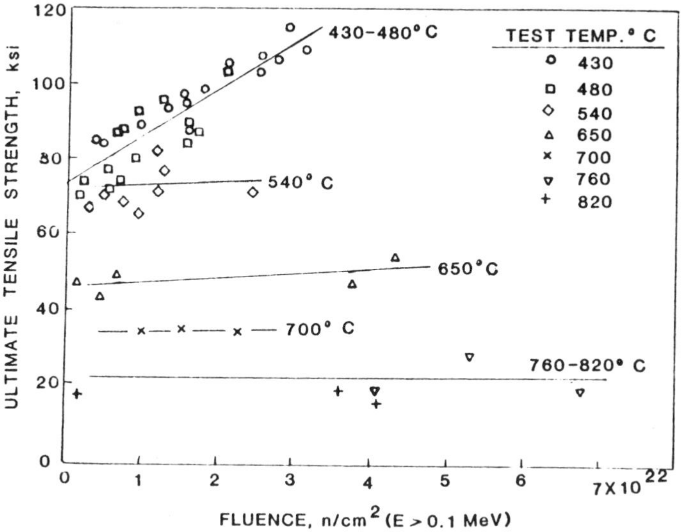

FIGURE 7.9 Effect of fluence on the ultimate tensile strength of Type 316 stainless steel.

[Reproduced with permission from R.L. Fish and J.J. Holmes, "Tensile Properties of Annealed Type 316 Stainless Steel After EBR-II Irradiation," J. Nucl. Materials, 46, 1973, pp. 113-120]

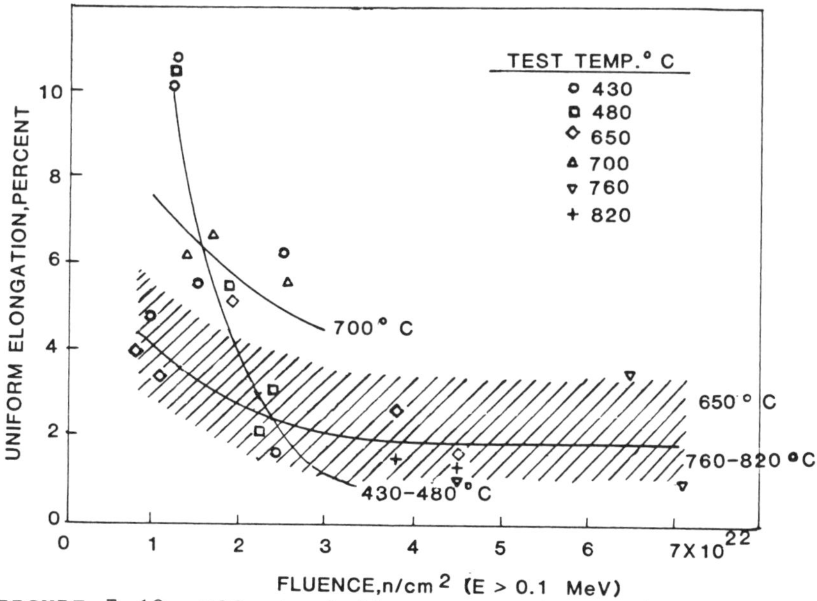

FIGURE 7.10 Effect of fluence on the uniform elongation of Type 316 stainless steel.

[From Fish and Holmes, Ref. (7-12)]

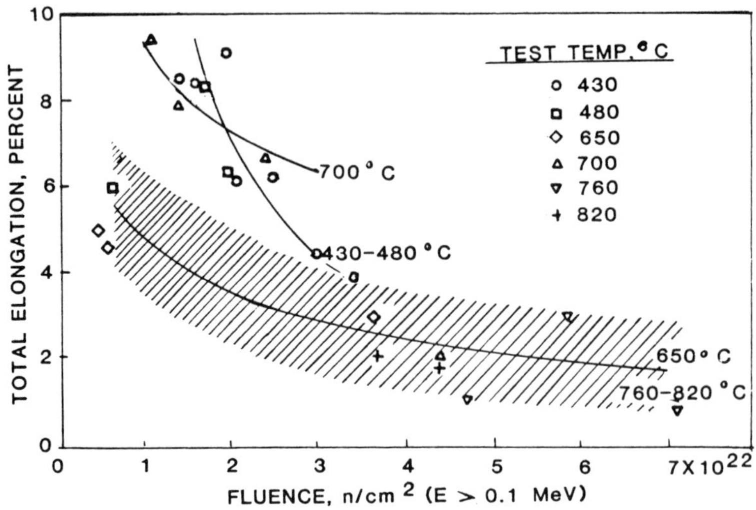

FIGURE 7.11 Effect of fluence on the total elongation of Type 316 stainless steel.
[From Fish and Holmes, Ref. (7-12)]

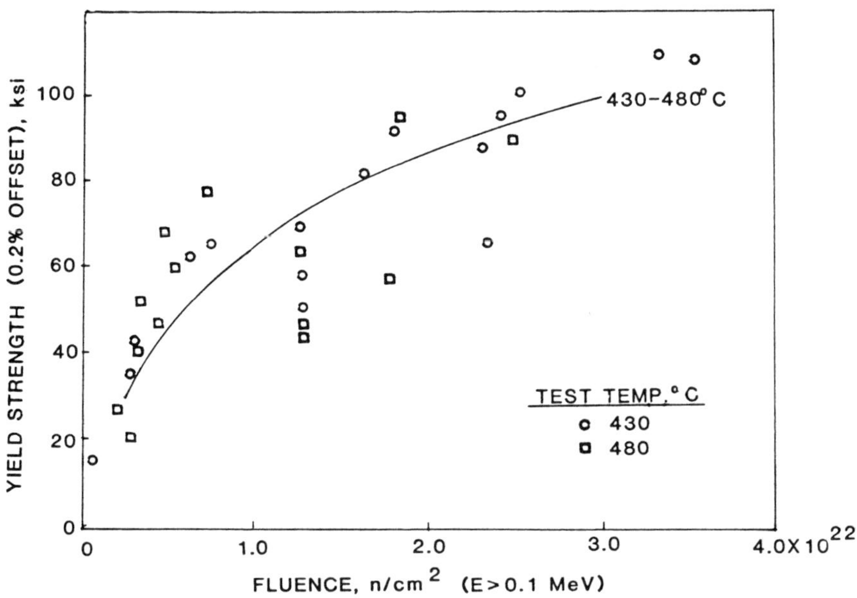

FIGURE 7.12 Effect of fluence on the low temperature (430-480°C) yield strength (0.2% offset) of Type 316 stainless steel.
[From Fish and Holmes, Ref. (7-12)]

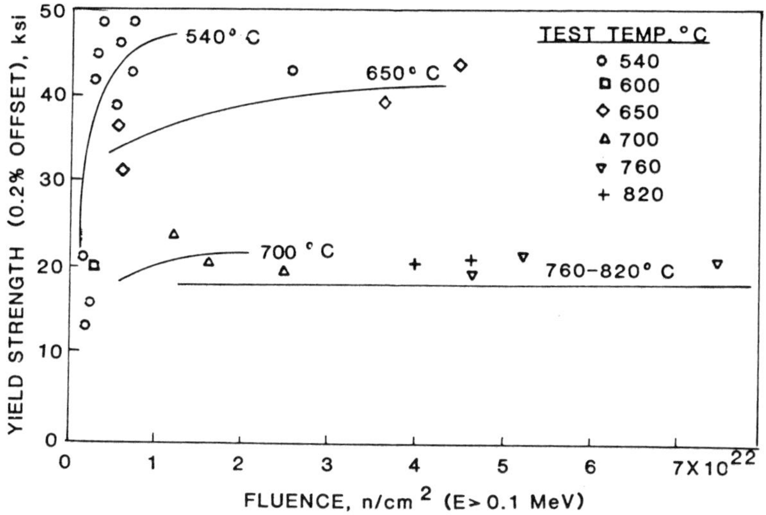

FIGURE 7.13 Effect of fluence on the high temperature (>540°C) yield strength (0.2% offset) of Type 316 stainless steel.
[From Fish and Holmes, Ref. (7-12)]

$$^{2}_{1}D + ^{3}_{1}T \rightarrow ^{4}_{2}He + ^{1}_{0}n \quad (7.13)$$

Studies show that helium causes blistering and spalling of the austenitic steel [Reference (7-14)]. Further, it induces elevated-temperature embrittlement [Reference (7-16)] and promotes swelling [Reference (7-17)].

7.1.4 Materials under cycle stress and strain. Many engineering components must withstand numerous load or stress reversals during their service life. A good example is the alternating stress associated with a rotating shaft. A detailed treatment of the effect of cyclic stress on fatigue life (indicated by the S-N curve) is given by Wohler [Reference (7-15)].

It is found that cyclic life increases with decreasing stress level and that below a certain stress level, it seems to have infinite life. Secondly the fatigue life is

Table 7.9

Tensile Properties of Annealed Type 316 Stainless Steel

Tensile properties of EBR-II irradiated annealed Type 316 stainless steel (strain rate = 1.78×10^{-3}/min)

Specimen Identity	Irradiation Temp. (°C)	Fluence (n/cm^2)		Strength (ksi)			Ductility (%)		
		(Total)	(E>0.1 MeV)	PEL	YS	UTS	UE	TE	RA
430°C Test Temperature									
Unirradiated Control	-	-	-	12.5	19.0	70.0	40.0	47.0	67.0
71126-1	460	2.3×10^{22}	1.3×10^{22}	41.5	51.4	80.7	13.2	15.9	56.1
A32-5T	420	2.0	1.3	53.3	62.2	90.5	9.8	13.0	53.8
A32-7T	410	3.2	2.5	76.9	98.5	102.5	1.7	5.0	48.8
A32-8T	400	3.8	3.2	74.2	110.1	112.1	1.1	4.2	43.0
A32-9T	390	4.0	3.4	87.6	108.5	109.7	0.9	3.9	50.3
6E16-1	430	2.7	1.86	65.0	87.8	95.9	4.8	8.0	57.7
6E26-1	430	3.25	2.4	71.1	92.9	100.9	3.1	6.3	48.0
7E10-2	400	3.3	2.3	66.6	87.8	98.8	6.9	9.1	55.3
4G16-1	390	2.3	1.3	52.0	69.0	87.0	10.1	12.9	60.0
4G26-1	390	2.7	1.6	53.0	79.0	91.3	5.2	8.1	59.3
16J16-2	420	0.5	0.23	25.2	35.0	76.5	17.9	20.7	63.4
16J26-1	420	0.4	0.18	16.3	25.9	77.0	23.5	26.5	63.4
19E16-1	460	1.15	0.75	45.5	61.0	83.7	12.0	16.2	60.1
7F16-1	470	3.3	2.3	48.8	64.2	87.0	10.5	12.7	56.1
19E26-1	460	0.95	0.58	43.0	58.3	86.2	16.3	20.0	65.0
480°C Test Temperature									
Unirradiated Control	-	-	-	12.5	18.5	69.0	42.0	48.0	66.0
14G26-3	510	0.35×10^{22}	0.19×10^{22}	16.7	17.6	69.7	41.3	51.9	65.0
15G16-14	510	0.25	0.14	20.5	24.7	72.0	42.0	52.0	65.3
6E16-2	430	2.7	1.86	60.2	87.8	95.1	4.9	8.0	52.0
6E26-2	430	3.3	2.4	67.2	88.0	92.8	2.2	5.1	48.4
5G16-2	440	2.3	1.3	52.0	66.0	83.3	9.9	12.2	52.0
7H26-3	460	2.3	1.3	30.0	51.0	80.7	14.1	16.2	50.4
4H16-1	460	2.7	1.6	34.2	53.7	79.7	11.9	14.3	48.8
4H26-1	460	2.3	1.3	44.8	49.0	79.8	13.2	14.1	49.6
19G16-1	490	0.75	0.4	46.3	54.9	82.5	37.7	40.9	57.7
19G26-1	490	0.6	0.3	30.1	49.8	80.5	52.4	63.7	60.2
21E16-2	480	1.15×10^{22}	0.75×10^{22}	45.5	82.1	91.0	4.7	6.3	43.9
21E26-1	480	0.95	0.58	58.5	72.4	87.4	13.3	15.9	49.6
14F26-9	500	0.4	0.25	24.4	32.5	71.4	29.5	33.0	65.9
14F16-15	500	0.5	0.33	30.9	39.8	73.2	26.5	30.2	64.2

Table 7.9 (Continued)

540°C Test Temperature									
Unirradiated Control	-	-	-	12.5	18.0	65.0	40.0	46.0	65.0
14G26-4	510	0.35 x 10^{22}	0.19 x 10^{22}	15.7	19.3	66.9	38.4	51.0	66.1
15G16-15	510	0.25	0.14	15.8	22.6	69.0	36.7	47.9	63.4
6F26-2	480	3.3	2.4	35.8	44.7	71.5	12.8	13.95	28.5
20F26-4	570	0.95	0.58	30.9	37.8	65.5	10.3	11.3	25.2
20F16-2	570	1.15	0.75	29.3	41.5	69.2	10.2	11.1	39.8
21H26-1	560	0.75	0.40	28.4	44.7	74.7	16.4	17.9	35.0
19F16-1	510	0.95	0.58	30.9	46.0	73.5	32.3	35.9	44.7
19F26-1	510	1.15	0.75	34.2	48.7	74.6	22.9	29.8	35.8
21H16-1	560	0.60	0.30	30.9	40.7	71.5	19.6	22.3	30.9
21F16-2	530	0.95	0.58	39.8	49.6	72.0	30.7	34.9	24.4
600°C Test Temperature									
Unirradiated Control	-	-	-	12.5	17.5	59.0	32.0	41.0	57.0
11H16-14	540	0.35 x 10^{22}	0.19 x 10^{22}	17.8	20.6	54.9	26.8	30.6	37.4
650°C Test Temperature									
Unirradiated Control	-	-	-	12.5	17.0	45.0	22.0	39.0	45.0
14B86-3	710	0.85 x 10^{22}	0.50 x 10^{22}	18.7	31.8	43.9	4.6	5.1	22.8
38A6-8	660	0.74	0.64	22.3	35.4	47.5	3.8	4.5	17.1
33-3-5	670	5.3	4.4	32.5	41.8	49.5	1.9	2.07	18.7
33-3-1	670	4.5	3.7	28.4	39.0	48.0	2.8	3.1	15.5
700°C Test Temperature									
Unirradiated Control	-	-	-	12.5	17.0	31.0	14.0	46.0	52.0
21B86-4	710	1.6 x 10^{22}	1.1 x 10^{22}	15.4	22.5	35.0	6.4	9.4	21.1
21B76-1	710	2.1	1.6	16.3	21.0	34.1	6.5	7.7	28.5
21B56-5	710	2.8	2.3	14.2	18.9	34.1	6.0	6.8	15.4
760°C Test Temperature									
Unirradiated Control	-	-	-	12.5	16.5	23.5	7.0	60.0	64.0
7B86-2	760	5.5 x 10^{22}	4.4 x 10^{22}	12.6	16.9	20.2	1.0	1.1	8.1
7B76-1	760	6.7	5.6	12.6	18.7	24.8	2.9	3.2	12.2
7B56-2	760	7.85	7.0	12.2	17.3	19.0	0.6	0.8	6.5
820°C Test Temperature									
Unirradiated Control	-	-	-	12.5	16.0	20.0	4.0	70.0	74.0
33-2-5	830	5.3 x 10^{22}	4.4 x 10^{22}	11.8	17.1	19.1	1.6[a]	2.0	8.1
33-2-6	830	4.5	3.7	13.0	18.7	20.2	1.5[a]	2.3	12.2

a) Evaluated directly from load-elongation chart.

PEL=proportional elastic limit YS=yield strength UTS=ultimate tensile strength UE=uniform elongation TE=total elongation RA=reduction in Ayca

[Reproduced with permission from R.L. Fish and J.J. Holmes, "Tensile Properties of Annealed Type 316 Stainless Steel After EBR-II Irradiation," J. Nucl. Materials, 46, 1973, pp. 113-120]

decreased by the presence of a notch, as already indicated on the previous S-N curve (Figure 7.6). Three stages are involved in fatigue: initiation, propagation and failure.

In the S-N curve (stress versus the number of cycles to failure), the stress parameter sometimes is expressed as the stress amplitude, which can be defined as below and is shown in Figure 7.14.

$$\sigma_a = \frac{\sigma_{max} - \sigma_{min}}{2} \qquad (7.14)$$

$$\sigma_m = \frac{\sigma_{max} + \sigma_{min}}{2} \qquad (7.15)$$

$$\Delta\sigma = \sigma_{max} - \sigma_{min} \qquad (7.16)$$

where σ_a = alternating stress amplitude, i.e., fatigue strength in terms of stress amplitude, $\sigma_m \neq 0$

σ_m = mean stress

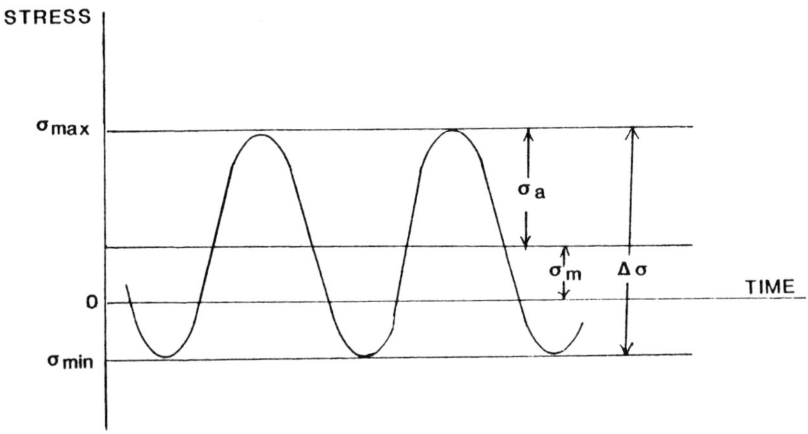

FIGURE 7.14 Nomenclature to describe test parameters involved in cyclic stress testing.

The effect of σ_m on fatigue life can be shown in Figure 7.15(a) [Reference (7-4)], i.e., an increasing σ_m will decrease cyclic life. Empirical formulae to account for the effect of mean stress on fatigue are as follows:

$$\text{Goodman relation } \sigma_a = \sigma_{fat}(1 - \frac{\sigma_m}{\sigma_{ts}}) \qquad (7.17)$$

$$\text{Gerber relation } \sigma_a = \sigma_{fat}[1 - (\frac{\sigma_m}{\sigma_{ts}})^2] \qquad (7.18)$$

$$\text{Soderberg relation } \sigma_a = \sigma_{fat}(1 - \frac{\sigma_m}{\sigma_{ys}}) \qquad (7.19)$$

where σ_{fat} = fatigue strength in terms of stress amplitude,

$\sigma_m = 0$ if $\sigma_{compressive} = \sigma_{tensile}$

σ_{ys} = yield strength

These three relations are shown in Figure 7.15(b), which shows the importance of σ_a and σ_m on fatigue endurance. Experience has shown that most data lie between the Gerber and Goodman diagram. The latter, then, represents a more conservative design criterion for mean stress effects.

A number of cumulative damage theories have been proposed during the last decade to describe the relative importance of stress interactions and the amount of damage (plastic deformation, crack initiation, and propagation) introduced to a component. Palmgren [Reference (7-18)] and Miner [Reference (7-19)] have developed the Palmgren-Miner cumulative damage law, which is simply as follows:

$$\sum_{i=1}^{k} \frac{n_i}{N_i} = 1 \qquad (7.20)$$

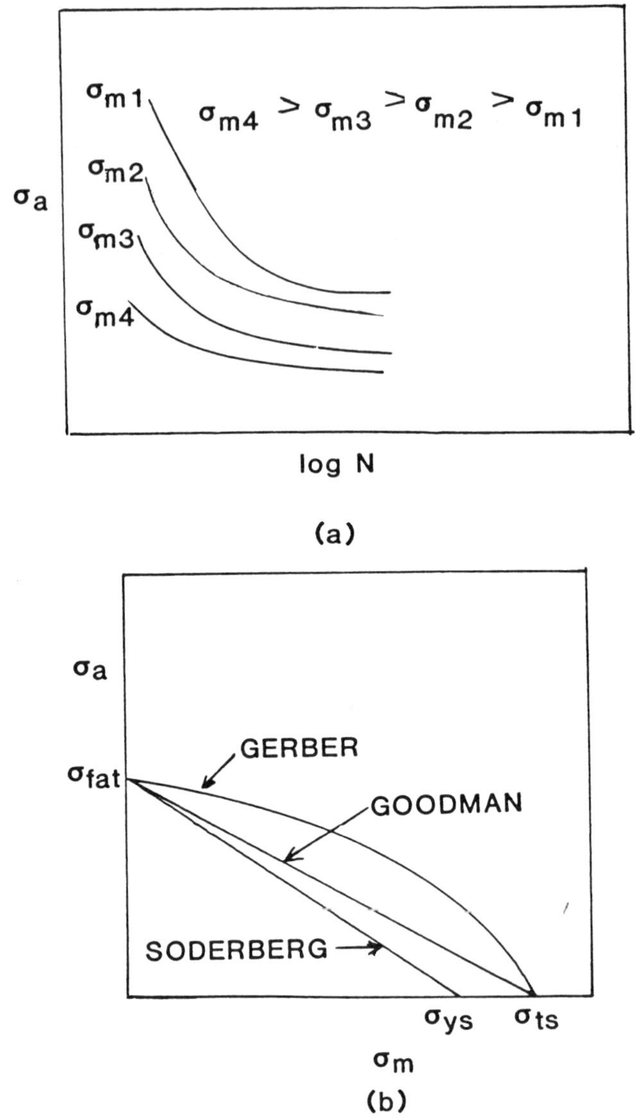

FIGURE 7.15 Representative plots of data showing effect of stress amplitude and mean stress on fatigue life. (a) Typical S-N diagrams with differing σ_m levels; (b) Gerber, Goodman, and Soderberg diagrams showing combined effect of alternating and mean stress on fatigue endurance.

[From Hertzberg, Ref. (7-4)]

where $i = i^{th}$ stress level

n_i = number of cycles applied at σ_i
N_i = fatigue life at σ_i
k = number of stress levels in the block loading spectrum.

Figure 7.16 shows the calculations of N_i for given σ_i such that

$$\frac{n_1}{N_1} + \frac{n_2}{N_2} + \frac{n_3}{N_3} = 1.$$

This can be generalized by using the cumulative damage law of Equation (7.20).

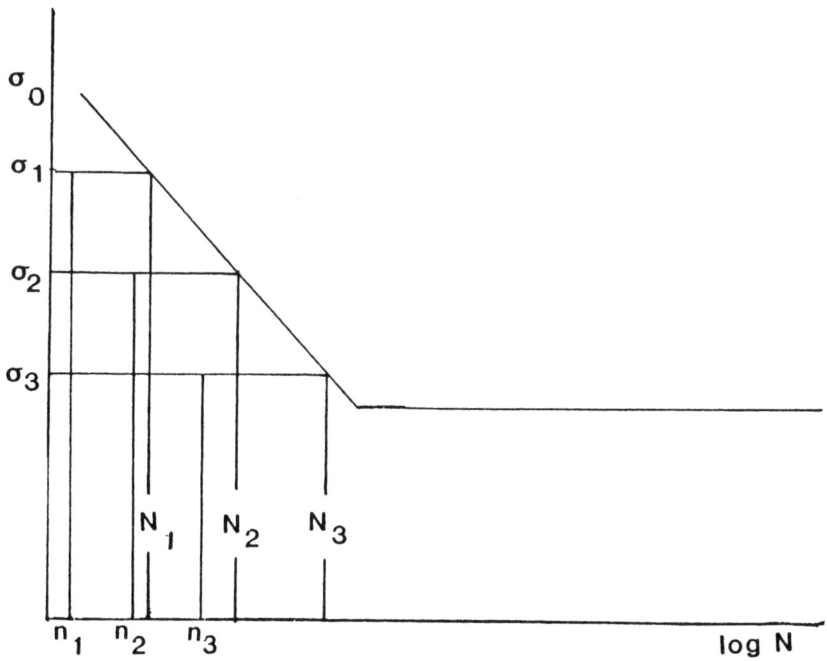

FIGURE 7.16 Component cyclic life determined from $\Sigma n_i/N_i = 1$ if damage at σ_i is linear function of n_i and damage is not a function of block sequencing.

In addition to stress-controlled fatigue, certain cases of strain-controlled fatigue are of interest. A good example of strain-controlled fatigue is thermal cycling, where components expand and contract throughout an operating range of temperature. For ideally elastic materials, the hysteresis loop for cyclic loading is shown in Figure 7.17 [Reference (7.14)]. For materials which exhibit elastic and plastic deformation, the loop is shown in Figure 7.18 [Reference (7-4)]. Now the total strain can be expressed as:

$$\Delta\varepsilon_T = \Delta\varepsilon_P + \Delta\varepsilon_e \tag{7.21}$$

where $\Delta\varepsilon_e = \frac{\Delta\sigma}{E}$ is the elastic strain range and $\Delta\varepsilon_p$ is the plastic strain range.

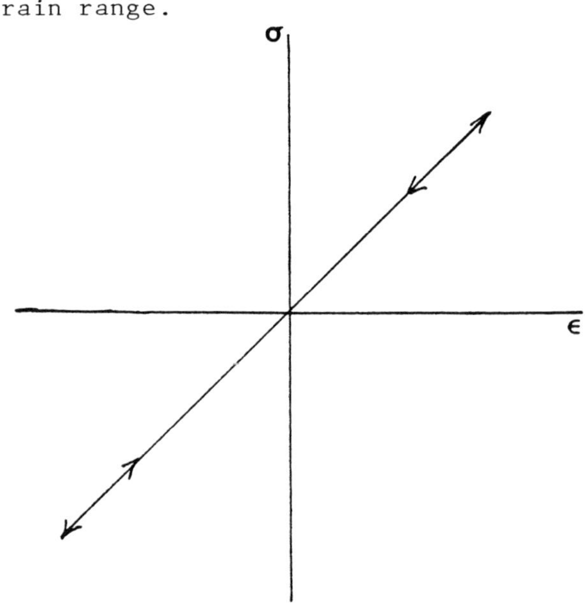

FIGURE 7.17 Hysteresis loop for cyclic loading in ideally elastic material.

[Reproduced with permission from S.K. Das, "The Role of Inert Gases in First Wall Phenomena in Fusion Devices," Proc. Harwell Consultants Symp. Inert Gases in Metals and Ionic Solids, AERE-R9733, Vol. II, p.409, S.F. Pugh, Ed., U.K. Atomic Energy Research Establishments, Mar. 1980]

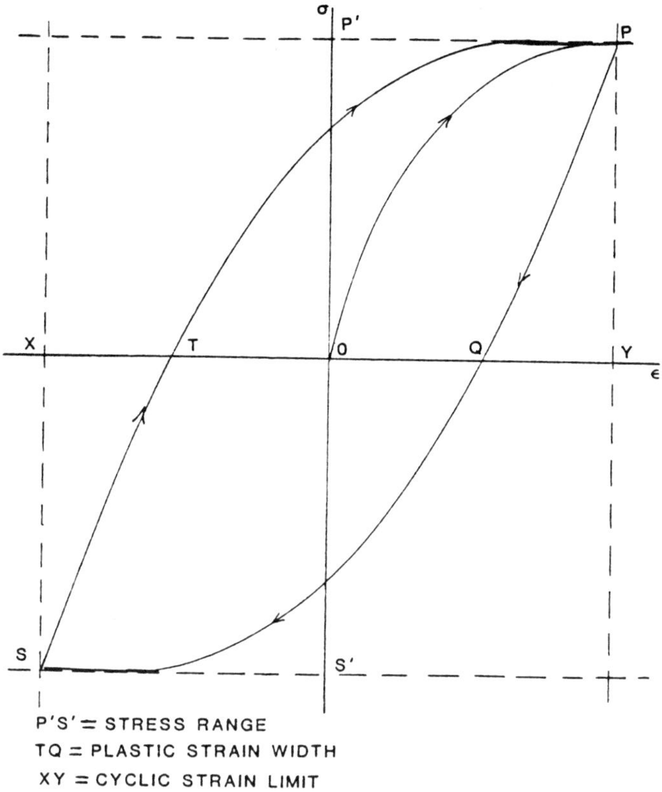

FIGURE 7.18 Hysteresis loop for cyclically loaded material undergoing elastic and plastic deformation.

[From Hertzberg, Ref. (7-4)]

Hence,

$$\Delta\varepsilon_p = \Delta\varepsilon_T - \frac{\Delta\sigma}{E} \tag{7.22}$$

In both cases of either stress-controlled or strain-controlled fatigue, the material either softens or hardens as shown in Figures 7.19 and 7.20 [Reference (7-20)]. As shown in these two figures, the material response changes with continuous cycling until cyclic stability is reached. The material becomes either more or less resistant to the applied stress or strains. Therefore, it is said to be cyclically strain hardened or strain softened. Manson, et al, [References (7-21) and (7-22)] observed

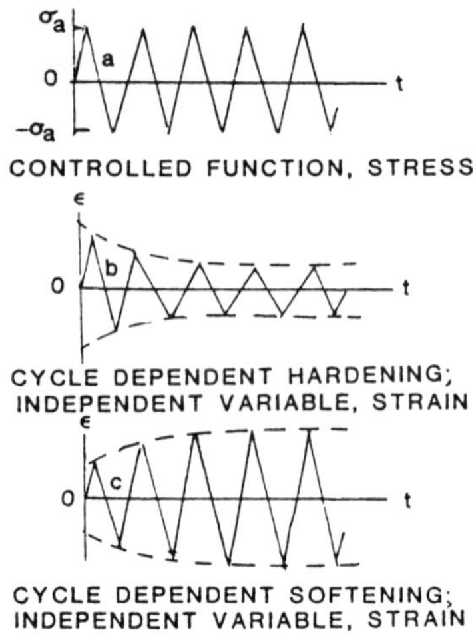

FIGURE 7.19 Cycle dependent material response under stress control.

[Reproduced with permission from B.I. Sandor, Fundamentals of Cyclic Stress and Strain, University of Wisconsin Press, Madison, Wis., 1972]

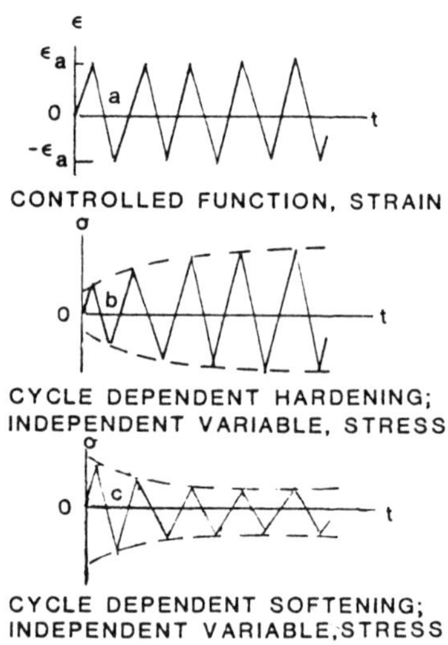

FIGURE 7.20 Cycle dependent material response under strain control.

[From Sandor, Ref. (7-20)]

that whether the material is cyclically hardened or softened depends upon the ratio of the monotonic ultimate strength (σ_{ult}) to the yield strength (σ_y) such that if $\frac{\sigma_{ult}}{\sigma_y} > 1.4$, the material will harden but if $\frac{\sigma_{ult}}{\sigma_y} < 1.2$, softening will occur.

The elastic component of the strain is often described in terms of the relationship between the true stress amplitude and number of load reversals.

$$\frac{\Delta\varepsilon_e E}{2} = \sigma_a = \sigma_f' (2N_f)^b \tag{7.23}$$

where

$\frac{\Delta\varepsilon_e}{2}$ = elastic strain amplitude
E = modulus of elasticity
σ_a = amplitude stress
σ_f' = fatigue strength coefficient, defined by the stress intercept at one load reversal ($2N_f=1$)
N_f = cycles to failure
$2N_f$ = number of load reversals to failure
b = fatigue strength exponent

Morrow [Reference (7-23)] determined that

$$b = \frac{-\bar{n}}{1+5\bar{n}} \tag{7.24}$$

where $\bar{n}$ = strain hardening factor for cyclic stress-strain diagram. Equation (7.23) describes the elastic component of the strain. The plastic component of strain is described by the Coffin-Manson relationship, [References (7-21), (7-22), (7-24), and (7-25)].

$$\frac{\Delta\varepsilon_p}{2} = \varepsilon_f' (2N_f)^c \tag{7.25}$$

where

$\frac{\Delta\varepsilon_p}{2}$ = plastic strain amplitude

ε_f' = fatigue ductility coefficient, defined by strain intercept at one load reversal ($2N_f=1$)

$2N_f$ = total number of strain reversals to failure

c = fatigue ductile exponent, which is a material property in the range -0.5 to -0.7

Morrow [Reference (7-23)] determined that

$$c = \frac{-1}{1+5\bar{n}} \tag{7.26}$$

The relation between the monotonic true fracture ductility, ε_f, and ε_f' is usually accepted as

$$\varepsilon_f' = 0.35\varepsilon_f \tag{7.27}$$

By combining Equations (7.22), (7.23) and (7.25), the total strain amplitude may be given by

$$\frac{\Delta\varepsilon_T}{2} = \frac{\Delta\varepsilon_e}{2} + \frac{\Delta\varepsilon_p}{2} = \frac{\sigma_f'}{E}(2N_f)^b + \varepsilon_f'(2N_f)^c \tag{7.28}$$

The total strain life curve is shown in Figures 7.21(a) and (b) [Reference (7-26)] which indicates that the total strain life will approach the plastic strain life at large total strain amplitude, and the elastic strain life at low total strain amplitude.

The fatigue properties of SAE 4340 steel are shown in Figures 7.22(a) and (b) [References (7-21) and (7-23)]. An estimate ranging from $\varepsilon_f' = 0.35\ \varepsilon_f$ to $1.0\ \varepsilon_f$ has been used to analyze fatigue data since the extrapolation of the curve in Figure 7.22(b) does not always coincide with the monotonic fracture strain as determined from a tensile test. The stress amplitude-fatigue life behavior of hardened steels is shown in Figure 7.23 [Reference (7-26)] which demonstrates increasing fatigue life with decreasing fatigue strength. As mentioned previously, for the

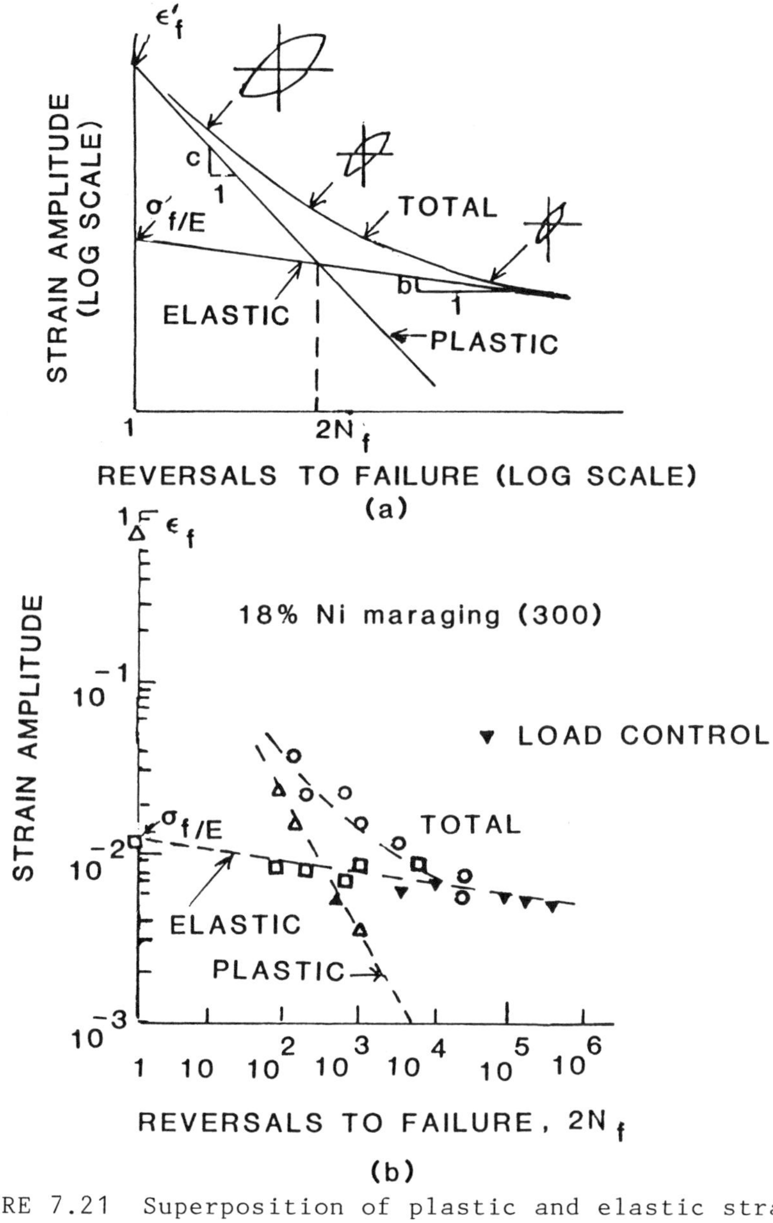

FIGURE 7.21 Superposition of plastic and elastic strain life curves to produce the total strain life fatigue relationship. (a) Schema; (b) actual data from 18% Ni maraging steel.

["Reprinted with permission from the American Society for Testing and Materials, 1916 Race Street, Philadelphia, PA 19103"; Landgraf, Ref. (7-26)]

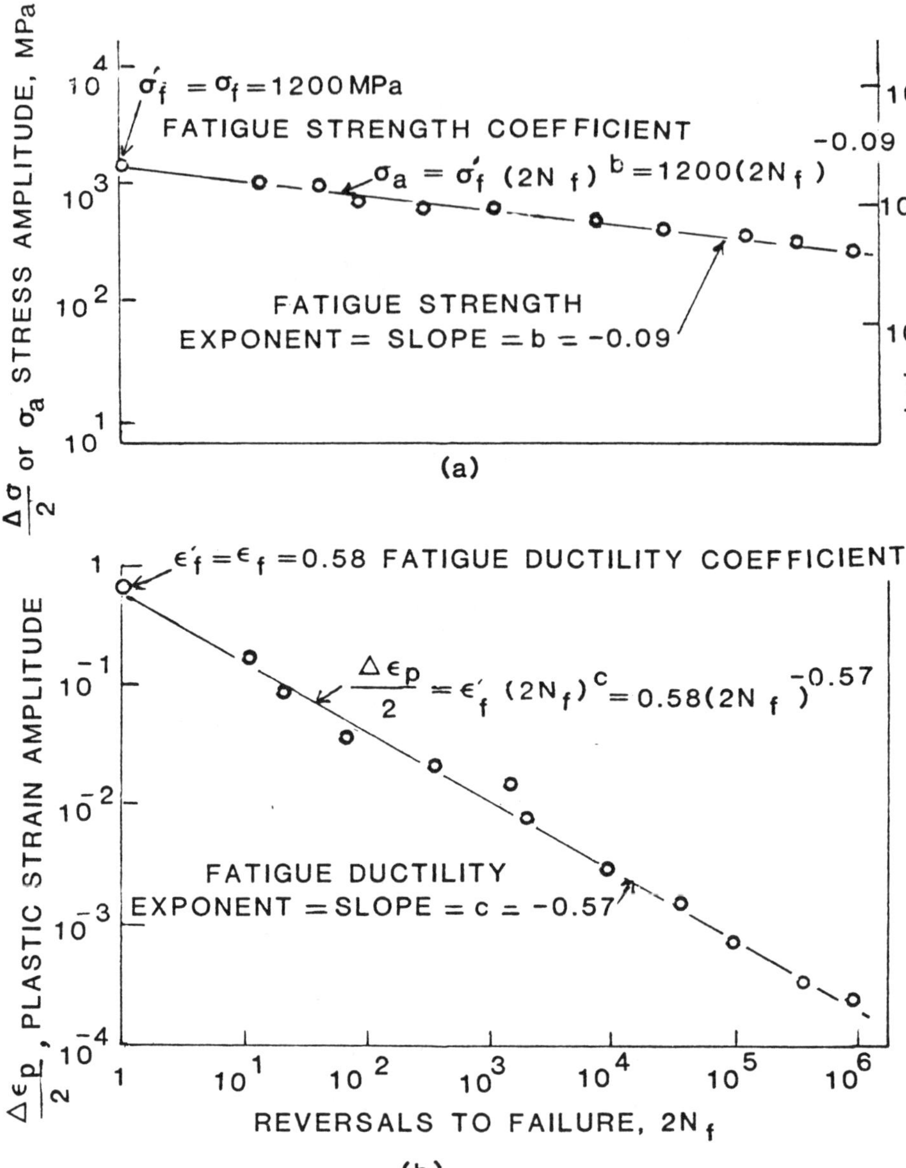

FIGURE 7.22 Fatigue properties of SAE 4340 steel. (a) Fatigue strength properties; (b) fatigue ductility properties.

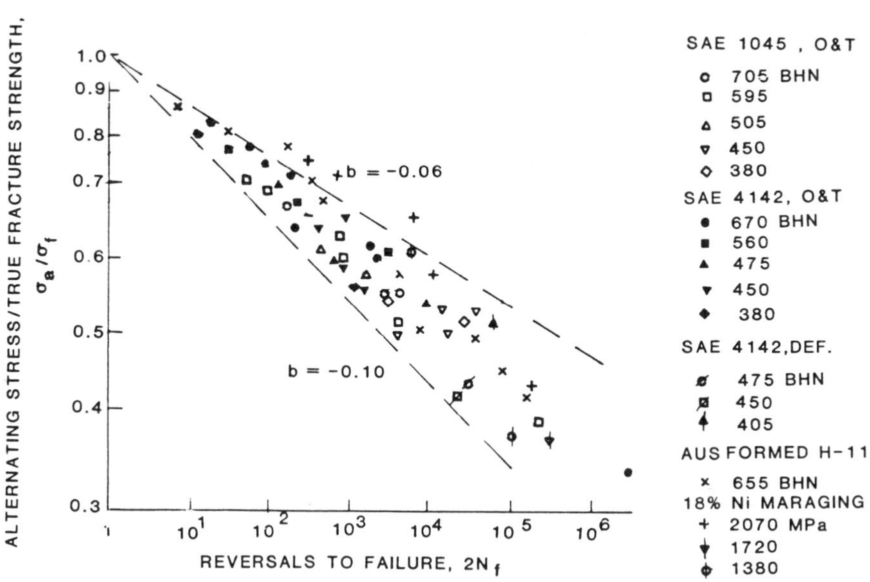

FIGURE 7.23 Stress amplitude-fatigue life behavior of hardened steels.
[From Landgraf, Ref. (7-26)]

exponent, based upon an energy argument, Morrow [Reference (7-23)] determined that

$$b = \frac{-\bar{n}}{1+5\bar{n}} \tag{7.24}$$

Consequently, fatigue life under essentially elastic stress (or strain) cycling is enhanced by a low value of $\bar{n}$. One would also want the material to have a high fatigue strength coefficient σ_f'. Many observations have shown that σ_f' is approximately equal to the monotonic fracture strength σ_f, so that a very strong alloy would be preferred for predominantly elastic strains associated with high cycle fatigue, as shown in Figure 7.24 [Reference (7-26)].

7.1.5 Radiation effects on fatigue stress. In cyclic loading there are many phenomena which are absent in a tensile test. Since the temperature in a fission or fusion reactor can be relatively high, there may be associated thermal fatigue. The differences between mechanical and thermal fatigue are as follows [Reference (7-27)].

1. In thermal stress fatigue, the plastic strain tends to become concentrated in the hottest region of the body.

2. In thermal fatigue, there is a localized accumulation of strain. In mechanical fatigue, at a constant elevated temperature, the same regions are alternately subjected to plastic flow in tension and compression.

3. Cyclic variation in temperature may, itself, have an important effect on the material.

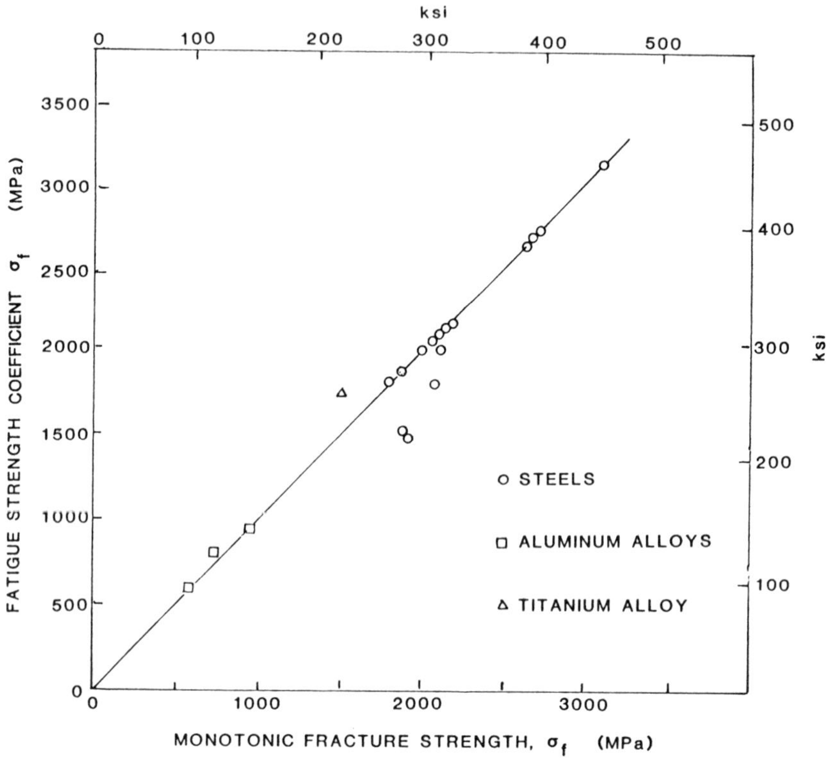

FIGURE 7.24 Correlation between fatigue strength coefficient and monotonic fracture strength. [From Landgraf, Ref. (7-26)]

For engineers the most important stesses are yield strength (σ_y) and ultimate tensile strength (σ_{ult}). The true stress-strain curves for Types 304 and 316 stainless steel are shown in Figures 7.25 and 7.26 [Reference (7-28)]. It is clear that the true stress at a given strain tends to decrease with an increase in test temperature. The curves also indicate that Type 316 stainless steel is stronger than Type 304 in the temperature range 550 °C to 750 °C.

Data are sparse on radiation effects on fatigue stress for stainless steel. One reason is that it is very difficult to investigate the plastic range in the equations which are used to describe fatigue properties. Fatigue life is generally ductility-dependent in the low-cycle range where plastic strain dominates and strength-dependent in the high-cycle range where elastic strain dominates. Several semiempirical expressions [References (7-29) and (7-30)] have been developed that incorporate this concept and relate tensile properties to fatigue life. Two of them are as follows [Reference (7-32)]:

a. Universal slopes equation:

$$\Delta\varepsilon_t = \frac{3.5\,\sigma_{ult}}{E}(N_f)^{-0.12} + D^{0.6}N_f^{-0.6} \qquad (7.29a)$$

b. Characteristic slopes equation:

$$\Delta\varepsilon_t = 2\varepsilon_{ef}\left(\frac{N_f}{10}\right)^{-m/2} + \frac{D^2}{\dot{\varepsilon}}\left(\frac{N_f}{f}\right)^{-1} \qquad (7.29b)$$

where

$\Delta\varepsilon_t$ = total strain rate

σ_{ult} = ultimate strength

E = Young's modulus of elasticity

D = fracture ductility = ln [100/(100-RA)]

RA = reduction in area, percent

N_f = number of cycles to failure

ε_{ef} = elastic strain rate at fracture

= true fracture stress divided by Young's modulus

m = strain-hardening coefficient

$\dot{\varepsilon}$ = strain rate = $2f\Delta\varepsilon_t$

f = cyclic frequency

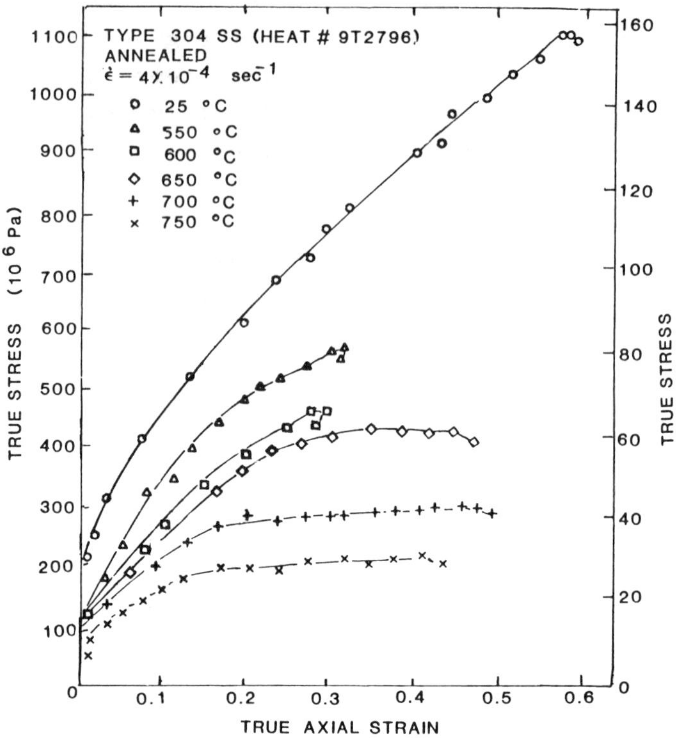

FIGURE 7.25 True stress-true strain behavior of Types 304 at various temperatures.

[Reproduced with permission from K. Natesan, et al, "In Symposium on Structural Materials for Service at Elevated Temperatures in Nuclear Power Generation," ASME Winter Meeting, Nov. 30 - Dec. 3, 1975, Houston, Texas, pp. 302-315]

A method has been proposed for estimating irradiated fatigue properties of materials from a knowledge of the unirradiated strain-controlled properties and the irradiated and unirradiated tensile properties [Reference (7-31)]. This method is based on a fraction modification of the irradiated fatigue data defined by making a fractional correction as follows:

$$\Delta\varepsilon_t = \Delta\varepsilon_e + \Delta\varepsilon_p = AN_f^a + BN_f^b \tag{7.30}$$

where

$\Delta\varepsilon_e$, $\Delta\varepsilon_p$ = elastic and plastic strain range respectively
A,B,a,b = constants for unirradiated material

Then for irradiated material Equation (7.30) is modified by

$$(\Delta\varepsilon_t)_{irrad} = \Phi_u A N_f^a + \Phi_t B N_f^b \qquad (7.31)$$

where

$$\Phi_u = \frac{(\sigma_{ult})_{irrad}}{(\sigma_{ult})_{unirrad}}$$

$$\Phi_t = \left(\frac{D_{irrad}}{D_{unirrad}}\right)^{0.6}$$

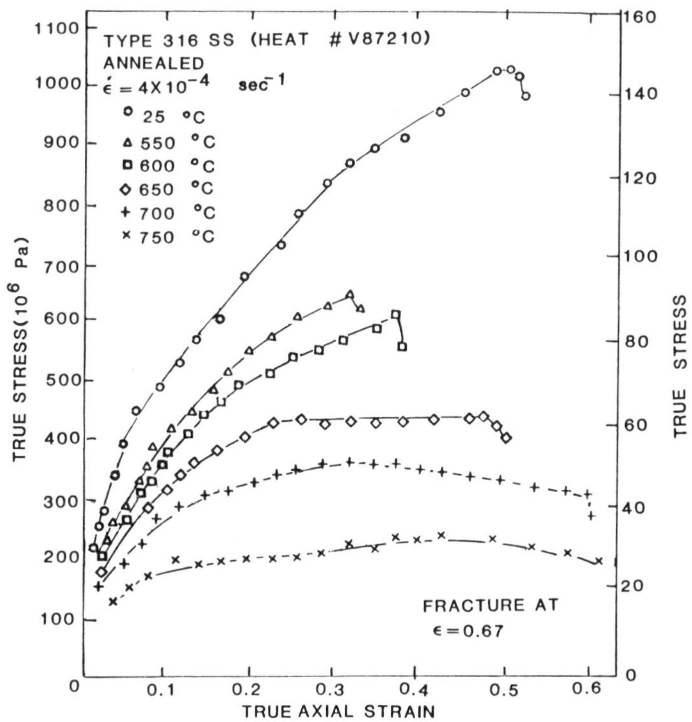

FIGURE 7.26 True stress-true strain behavior of Types 316 at various temperatures.
[From Natesan, Ref. (7-28)]

The method of characteristic slopes gives nonconservative estimates of fatigue lifetimes for the unirradiated material [Reference (7-32)], but provides either a conservative estimate or good fit of data for irradiated material. The method of universal slopes gives a conservative estimate for both the irradiated and unirradiated material in all cases. The fraction modification method was found to provide an excellent estimate of the fatigue behavior of the irradiated stainless steel. Beeston and Brinkman [Reference (7-33)] showed that reduction in fatigue life attributable to irradiation ranged from a factor of 1.5 to 2.5 for stainless steel. Data given in Tables 7.10 through 7.12 and in Figures 7.27 through 7.29 show that no difference in the fatigue life of Types 304 and 316 stainless steel is evident [Reference (7-32)]. Similar results were found in comparing the fatigue behavior of Types 304 and 304L stainless steel irradiated to a neutron fluence of 2 to 5 x 10^{21} n/cm^2 (E_n>0.1 MeV) at 750 °C and tested at 500 °C. The following was concluded for Types 304, 304L, and 316 stainless steel:

1. Irradiation to a neutron fluence of 0.4 to 5 x 10^{22} n/cm^2 (E_n >0.1 MeV) with temperature ranging from 700 °C to 750 °C resulted in a decrease in the tensile ductility.

2. Decrease in fatigue lifetime is proportional to the reduction in ductility.

3. Reduction in fatigue life ranges from a factor of 1.5 to 2.5.

Beeston and Brinkman have found the following further conclusions for Types 304, 304L, and 316 stainless steel [Reference (7-33)]:

1. Irradiation at temperatures of 450°C or 750°C to a

TABLE 7.10

Fatigue Properties of Type 304 Stainless Steel, Heat 55697, Tested at 700°C (1292°F)

Specimen No.	Aging or Irradiation Temperature, deg°C	Fluence, n/cm^2 E>0.1 MeV	Strain Rate $\dot{\varepsilon}$, s^{-1}	Axial Strain Range		Stress		Cycles to Failure, N_f	Time to Failure, t_f, h
				Total $\Delta\varepsilon_t$, %	Plastic $\Delta\varepsilon_p$, %	Range, $\Delta\sigma$, kg/mm^2(ksi)	Max Tensile, σ_t, kg/mm^2(ksi)		
30-E	750[a]	0	8×10^{-4}	0.35	0.14	31.1 (44.3)	15.6 (22.2)	18 964	46.09
19-E	750	0	8	0.55	0.31	38.2 (54.3)	17.1 (24.4)	3 965	15.14
18-E	750	0	8	1.01	0.70	45.8 (65.1)	22.1 (31.5)	882	6.19
20-E	750	0	8	2.41	2.05	55.7 (77.8)	26.1 (37.1)	245	4.10
52-E	700 to 750[b]	3.0×10^{22}	8	0.35	0.15	30.2 (42.9)	14.7 (20.9)	14 108	34.29
57-E	700 to 750	1.7	8	0.57	0.32	37.7 (53.7)	17.4 (24.7)	2 453	9.71
53-E	700 to 750	1.5	8	1.04	0.72	41.8 (59.5)	19.6 (27.9)	520	3.76
55-E	700 to 750	2.9	8	2.45	2.09	54.8 (77.9)	26.1 (37.1)	22	0.37

a Thermal controls were aged in argon for 9240 h at temperature shown.
b Irradiated specimens were sealed in argon-bonded capsules during irradiation.

TABLE 7.11

Fatigue Properties of Type 304L (Ti modified) Stainless Steel, Heat 3756, Tested at 700°C (1292°F)

Specimen No.	Aging or Irradiation Temperature, deg°C	Fluence, n/cm^2 E>0.1 MeV	Strain Rate $\dot{\epsilon}$, s^{-1}	Axial Strain Range: Total $\Delta\epsilon_t$, %	Axial Strain Range: Plastic $\Delta\epsilon_p$, %	Stress: Range, $\Delta\sigma$, kg/mm^2(ksi)	Stress: Max Tensile, σ_t, kg/mm^2(ksi)	Stress: Tensile at End of Hold Period, σ_t, min kg/mm^2(ksi)
25-D	750 C[a]	0	3 X 10^{-4}	0.54	0.26	42.0 (59.7)	19.5 (27.8)	--
23-D	750	0	3	1.01	0.68	49.7 (70.7)	22.3 (31.7)	--
24-D	750	0	3	2.40	2.00	60.1 (85.5)	29.3 (41.7)	--
D-99	600[b]	0	30	0.61	0.31	45.3 (64.5)	22.1 (31.4)	--
D-98	600	0	30	2.41	2.07	62.9 (89.5)	30.0 (42.7)	--
D-42	750	0	8	0.59	0.39	37.0 (52.6)	17.5 (24.9)	13.1 (18.6)
D-41	750	0	8	0.57	0.41	34.8 (49.5)	15.6 (22.2)	6.6 (9.4)
D-90	600	0	8	0.60	0.42	38.9 (55.4)	19.4 (27.6)	10.3 (14.7)
30-D	750	0	8	2.38	2.21	53.2 (75.7)	25.2 (35.8)	9.2 (13.1)
D-91	600	0	8	1.0	0.76	46.0 (65.4)	22.3 (31.7)	13.0 (18.5)
D-93	600	0	8	2.35	2.13	61.5 (87.5)	31.3 (44.6)	11.9 (16.9)
D-92	600	0	8	2.34	2.15	55.5 (79.0)	28.0 (39.8)	8.0 (11.4)
64-D	700 to 750[c]	2.1 X 10^{22}	8	0.36	0.12	36.9 (52.5)	18.0 (25.6)	--
70-D	700 to 750	1.3	8	0.58	0.15	51.8 (73.7)	25.7 (36.6)	--
62-D	700 to 750	1.8	8	1.05	0.72	49.9 (71.0)	24.5 (34.9)	--
63-D	700 to 750	2.4	8	2.44	2.04	60.0 (85.3)	29.0 (41.2)	--
69-D	700 to 750	2.0	8	0.58	0.44	37.2 (52.9)	18.6 (26.4)	13.2 (18.8)
68-D	700 to 750	3.7	8	0.58	0.45	33.6 (47.8)	16.2 (23.1)	9.8 (13.9)
D-16	700 to 750	1.5	8	0.59	0.40	35.8 (50.9)	17.8 (25.3)	8.5 (12.1)
65-D	700 to 750	2.3	8	2.29	2.05	53.0 (75.4)	25.4 (36.2)	9.6 (13.7)
D-20	700 to 750	1.1	8	2.35	2.14	58.3 (82.9)	29.4 (41.8)	9.0 (12.8)

[a] All 750°C thermal controls aged in argon for 9240 h.
[b] All 600°C thermal controls aged in argon for 3096 h.
[c] Irradiated specimens were sealed in argon-bonded capsules during irradiation.

[From Brinkman, Ref. (7-32)]

Table 7.12

Fatigue and Creep/Fatigue Properties of Type 316 Stainless Steel, Heat 65808, Tested at 700°C (1292°F)

Specimen No.	Aging or Irradiation Temperature, deg C	Fluence, n/cm E>0.1 MeV	Strain Rate $\dot{\epsilon}$, s^{-1}	Axial Strain Range		Stress		Cycles to Failure, N_f	Time to Failure, t_f, h
				Total $\Delta\epsilon_t$, %	Plastic $\Delta\epsilon_p$, %	Range, $\Delta\sigma$, kg/mm^2(ksi)	Max Tensile, σ_t, kg/mm^2)ksi)		
0-G	750[a]	0	8×10^{-4}	0.32	0.15	26.4 (37.5)	12.5 (17.8)	29 265	65.03
14-G	750[b]	0	8	0.33	0.15	28.1 (40.0)	12.6 (18.0)	18 906	43.33
17-G	750[b]	0	0.6	0.32	0.17	22.5 (32.0)	9.8 (13.9)	3 721	110.25
G-37	750[b]	0	8	0.91	0.68	35.9 (51.1)	15.1 (21.5)	1 256	7.94
3-G	750[a]	0	0.6	0.95	0.82	21.5 (30.6)	11.0 (15.6)	426	37.47
18-G	750[b]	0	0.6	0.96	0.81	23.1 (32.8)	11.5 (16.3)	486	43.20
8-G	--[c]	0	8	0.97	0.73	37.5 (53.4)	19.4 (27.6)	522	3.52
15-G	750[b]	0	8	0.98	0.74	37.3 (53.0)	18.1 (25.8)	408	2.78
9-G	--[c]	0	8	2.39	2.12	41.3 (58.8)	19.8 (28.1)	117	1.94
23-G	750[a]	0	0.6	2.39	2.19	30.7 (43.7)	15.0 (21.4)	195	43.15
21-G	750[a]	0	8	2.38	2.15	43.0 (61.2)	20.3 (28.9)	91	1.50
13-G	750[b]	0	8	2.34	2.02	35.4 (50.3)	15.7 (22.3)	132	2.14
44-G	700 to 750[d]	4.7×10^{22}	8	0.28	0.10	26.7 (38.0)	12.2 (17.4)	507	3.52
45-G	700 to 750[d]	4.0	0.6	0.31	0.19	20.1 (28.6)	9.3 (13.3)	310	8.90
G-1	700 to 750[d]	0.42	8	0.31	0.20	24.0 (34.1)	11.4 (16.2)	8 325	17.92
34-G	700 to 750[d]	3.4	8	0.32	0.15	27.4 (39.0)	12.4 (17.7)	1 043	2.32
42-G	700 to 750[d]	2.1	0.6	0.97	0.86	17.5 (24.9)	8.5 (12.1)	47	4.22
55-G	700 to 750[d]	0.44	0.6	0.96	0.82	21.9 (31.1)	10.1 (14.4)	91	8.09
G-3	700 to 750[d]	0.94	8	0.98	0.81	26.6 (37.8)	12.6 (17.9)	85	0.58
40-G	700 to 750[d]	1.4	8	0.99	0.79	31.1 (44.3)	15.2 (21.7)	83	0.57
39-G	700 to 750[d]	2.9	8	1.00	0.75	38.2 (54.4)	18.2 (25.9)	42	0.29
50-G	700 to 750[d]	0.53	0.6	2.36	2.18	26.9 (38.3)	12.4 (17.6)	11	2.40
56-G	700 to 750[d]	0.29	8	2.37	2.24	35.4 (50.3)	16.2 (23.1)	16	0.26

a Thermal controls aged in argon for 1541 h.
b Thermal controls aged in argon for 9340 h.
c Annealed.
d Irradiated in argon-bonded capsules.

Table 7.12 (Continued)

Specimen No.	Cycles to Failure, N_f	Time to Failure t_f, h	Tensile Hold Time t_f, h	Creep Damage: During Each Hold Period, (t/t_r) R	Creep Damage: During Each Fatigue Cycle, (t/t_r) F	Creep Damage: Total $\Sigma(t/t_r)$ R + $\Sigma(t/t_r)$ F	Fatigue Damage, N/N_f	Total Damage
25-D	4066	15.25	0	0	4.60×10^{-5}	0.187	1	1.187
23-D	1408	9.87	0	0	--	--	--	--
24-D	153	2.55	0	0	5.81×10^{-3}	0.889	0.732	1.621
D-99	3928	1.56	0	0	1.03×10^{-4}	0.040	1	1.040
D-98	209	0.36	0	0	8.16×10^{-4}	0.170	1	1.170
D-42	2500	35.2	0.01	4.75×10^{-5}	4.60×10^{-5}	0.234	0.615	0.849
D-41	518[d]	53.8[d]	0.1	1.71×10^{-5}	4.60×10^{-5}	0.033[d]	0.127[d]	0.160[d]
D-90	262[d]	257.1[d]	1.0	1.27×10^{-3}	4.60×10^{-5}	--	--	--
30-D	57	6.65	0.1	7.48×10^{-4}	5.81×10^{-3}	0.374	0.273	0.647
D-91	530	55.08	0.1	8.92×10^{-4}		--	--	--
D-93	47[d]	46.05[d]	1.0	1.22×10^{-2}	5.81×10^{-3}	0.846[d]	0.225[d]	1.071[d]
D-92	23[d]	104.66[d]	5.0	2.94×10^{-3}	5.81×10^{-3}	--	--	--
64-D	36.654[d]	91.63[d]	0	0	0	0	1	1
70-D	1730	6.97	0	0	1.46×10^{-4}	0.253	1	1.253
62-D	526	3.83	0	0	--	--	--	--
63-D	86	1.46	0	0	1.85×10^{-2}	1.591	0.935	2.526
69-D	374	5.24	0.01	1.74×10^{-4}	1.46×10^{-4}	0.120	0.216	0.336
68-D	71	71.28	1.0	1.03×10^{-3}	1.46×10^{-4}	0.083	0.041	0.124
D-16	116	116.80	1.0	7.41×10^{-4}	1.46×10^{-4}	0.103	0.067	0.170
65-D	21	2.44	0.1	2.97×10^{-3}	1.85×10^{-2}	0.451	0.228	0.679
D-20	12	59.45	5.0	1.22×10^{-2}	1.85×10^{-2}	0.368	0.130	0.498

d Test stopped before failure occurred.

[From Brinkman, Ref. (7-32)]

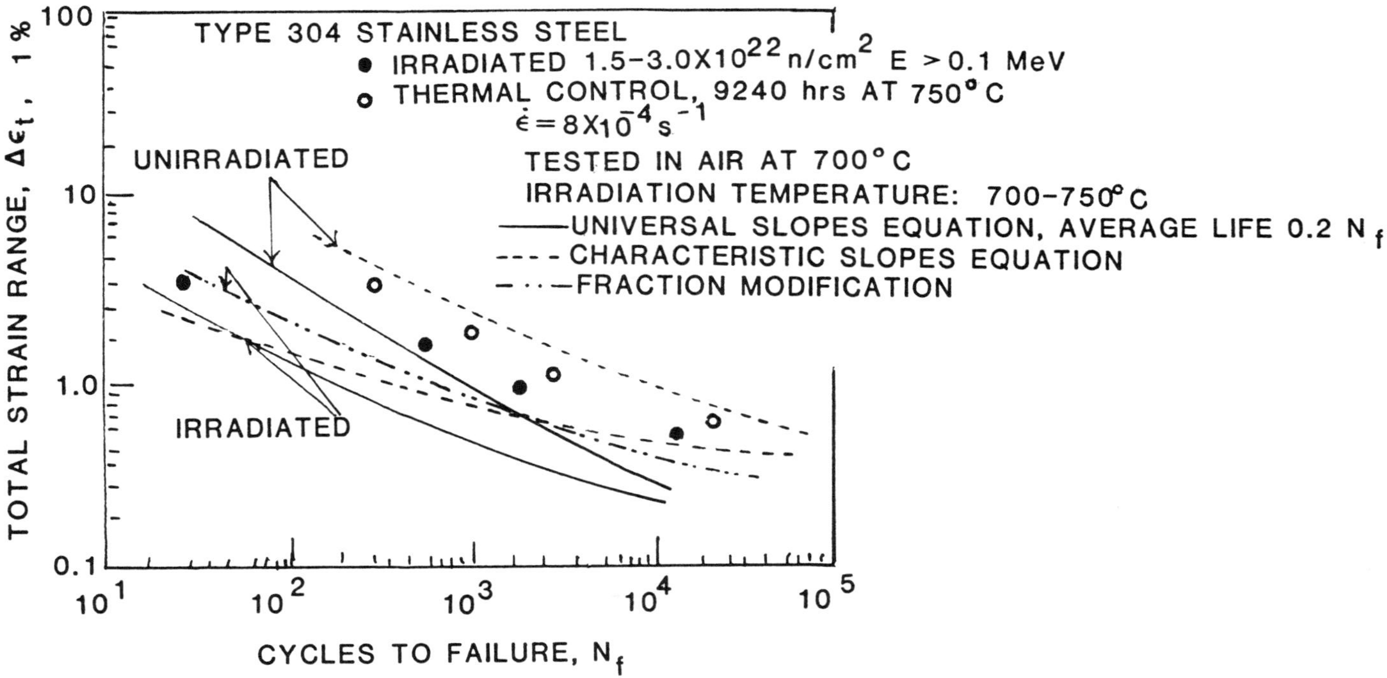

FIGURE 7.27 Irradiated and unirradiated fatigue behavior of Type 304 stainless steel tested at 700°C (1292°F) compared with predictions based on tensile data.

[From Brinkman, Ref. (7-32)]

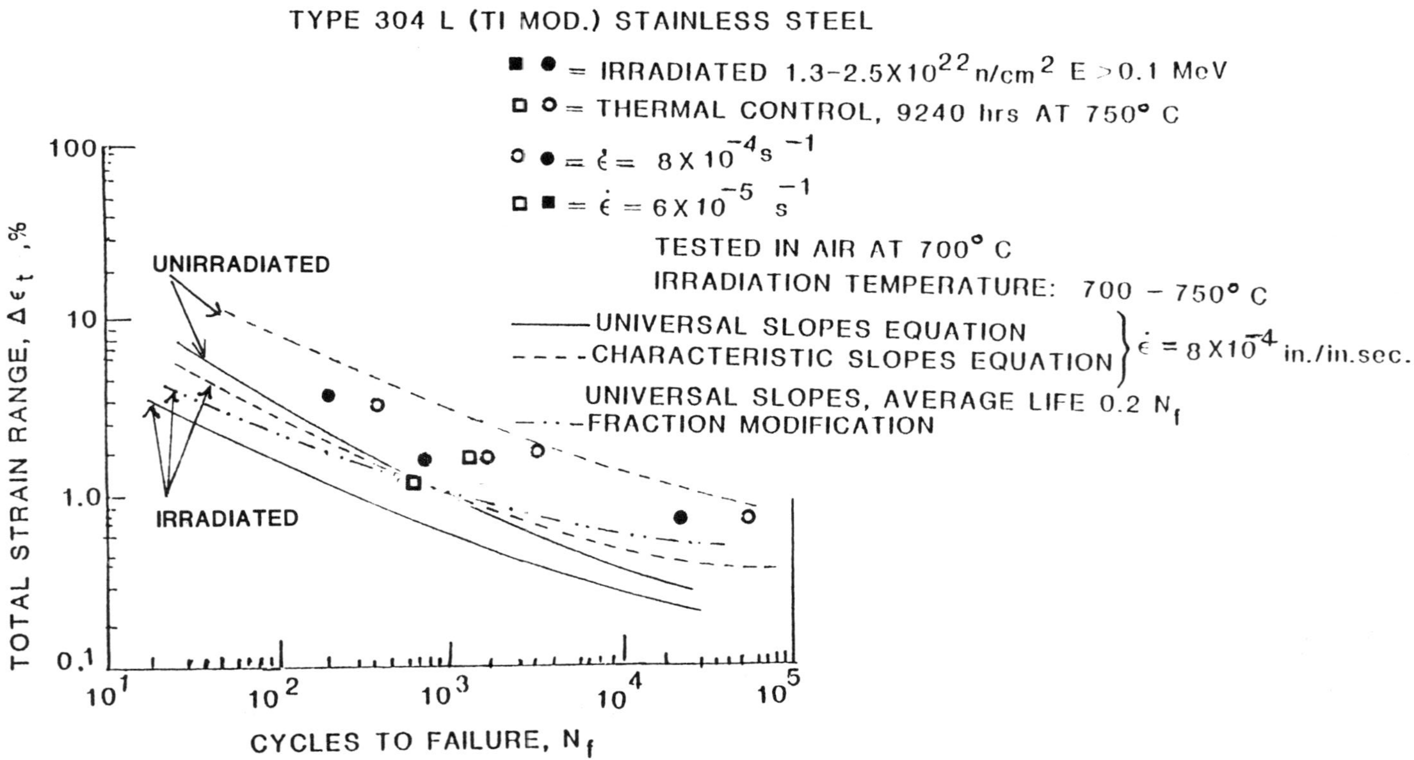

FIGURE 7.28 Irradiated and unirradiated fatigue behavior of Type 304L (titanium modified) stainless steel tested at 700°C (1292°F) compared with predictions based on tensile data. [From Brinkman, Ref. (7-32)]

neutron fluence less than 10^{22} n/cm^2 ($E_n > 0.1$ MeV) resulted in reduction in fatigue life by a factor of 2.5, which was attributed to an irradiation produced defect.

2. Changes in fracture ductility are indicated to be a dominant factor in the decrease in fatigue life after irradiation.

3. Generally, a given strain range required the stress range to be greater in the irradiated material, which is indicative of small yield stress.

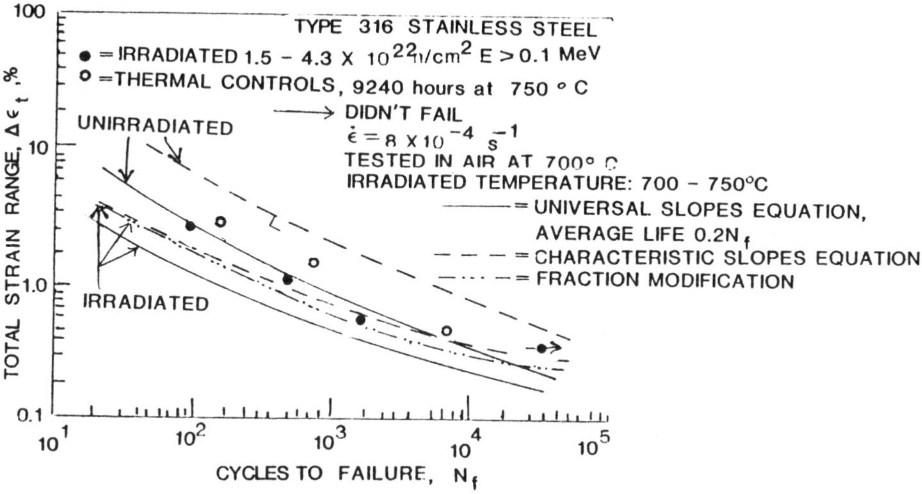

FIGURE 7.29 Irradiated and unirradiated fatigue behavior of Type 316 stainless steel tested at 700 °C (1292 °F) compared with predictions based on tensile data.

[From Brinkman, Ref. (7-32)]

Cyclic fatigue tests were performed by Battelle-Columbus [References (7-34) and (7-35)] on irradiated 347 stainless steel at 650 °C. The stainless steel was irradiated at 650 °C to a fast neutron fluence of 2.2 x 10^{22} n/cm^2 in the ETR and the results, summarized in Table 7.13, indicate that irradiation causes significant reductions in high temperature fatigue life. The reduction in fatigue life was attributed to the radiation-induced loss of ductility.

Cyclic strain fatigue tests on Type 316 stainless steel at 650°C by Berling and Conway [Reference (7-36)] demonstrated that irradiation causes significant reduction in fatigue life, even for specimens irradiated in the Oak Ridge Research Reactor at 70°C to a fast fluence of only 4.8×10^{18} n/cm^2. Excellent agreement was obtained by correlating the fatigue data for the irradiated specimens with the predicted curve obtained using the Berling-Conway equation [Reference (7-37)]. Figure 7.30 shows a comparison of fatigue data.

TABLE 7.13

Cyclic Fatigue Properties of Irradiated and Unirradiated 347 Stainless Steel Tested at 650°C

Stress Range, 1000 psi		Number of Cycles to Failure*	
Minimum	Maximum	Irradiated+	Unirradiated
2.0	22.0	2,368	15,324±
2.0	27.0	1,496	7,485

* The specimens were cycled at 50 cycles/hr using a cross-head speed of 0.02 in./min.

\+ Specimens were irradiated to a fast fluence of 2.2×10^{22} neutrons/cm^2.

± Specimen did not fail; test was discontinued.

[Taken from D.L. Keller (comp), "Progress on Development of Materials and Technology for Advanced Reactors During January through March 1969," USAEC Report BMI-1862, Battelle-Columbus, April 1969]

Battelle-Northwest workers performed irradiation tests for Type 316 stainless steel, showing that at elevated temperature, ductility is significantly reduced by irradiation. Both strength and ductility of the irradiated Type 316 stainless steel are highly sensitive to strain rate during testing [Reference (7-38)].

Argonne National Laboratory works on Type 304 stainless steel [Reference (7-39)] show that the irradiation temperature appears to be important, with more hardening resulting from irradiation at 370°C than those from the same fluence at 480°C.

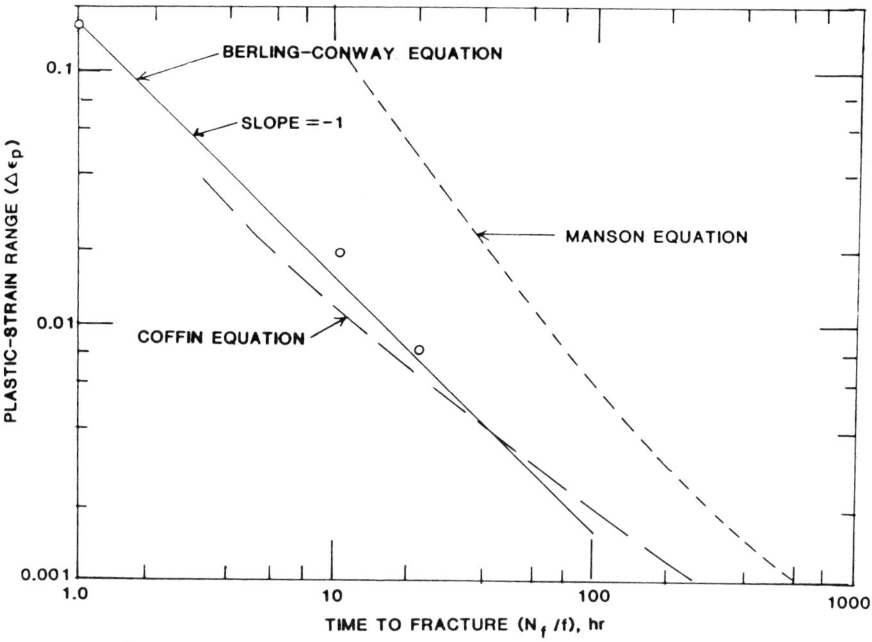

FIGURE 7.30 Comparison of the effectiveness of various equations for predicting fatigue behavior of irradiated 316 stainless steel tested at 650°C and a strain rate of 4×10^{-5} sec^{-1}.

[Reproduced with permission from J.T. Berling, and J.B. Conway, "A Proposed Method for Predicting the Low-Cycle Fatigue Behavior of 304 and 316 Stainless Steel," Trans.Met.Soc.AIME, May 1969, pp.1137-1140]

Martin and Weir [Reference (7-40)] studied the influence of irradiation temperatures from 150°C to 750°C, and post-irradiation test temperatures of 25°C to 850°C on mechanical properties of AISI Types 304 and 347 stainless steels tested at different strain rates. At lower irradiation and testing temperature their results indicate that both the yield and ultimate tensile stresses are strongly dependent upon irradiation temperature. Below 600°C test temperature the true tensile and fracture stresses were unaffected regardless of the irradiation temperature. At higher testing temperatures the radiation effect on the ductility parameters was strongly strain-rate sensitive. The effect of irradiation on mechanical properties of annealed Type 304 and 347 stainless steels is shown in Figures 7.31 through 7.33, and a summary of the results is shown in Table 7.14 [Reference (7-40)].

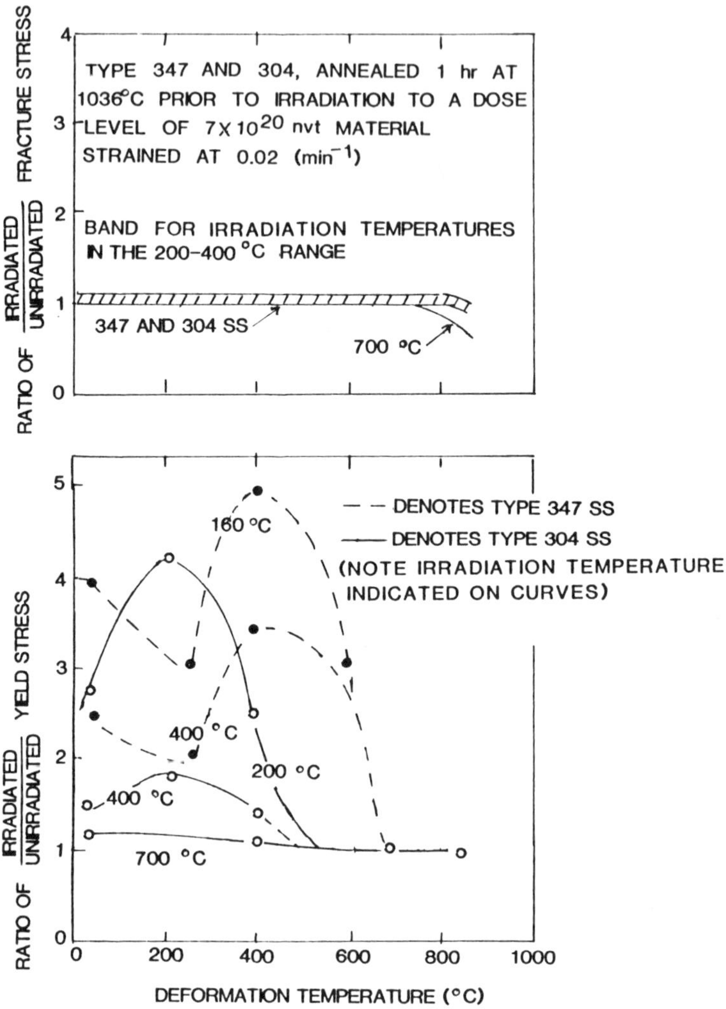

FIGURE 7.31 Effect of irradiation on the yield and fracture stress of annealed Type 304 and 347 stainless steels as a function of deformation temperature.

["Reprinted with permission from the American Society for Testing and Materials, 1916 Race Street, Philadelphia, PA 19103"; Martin, Ref. (7-40)]

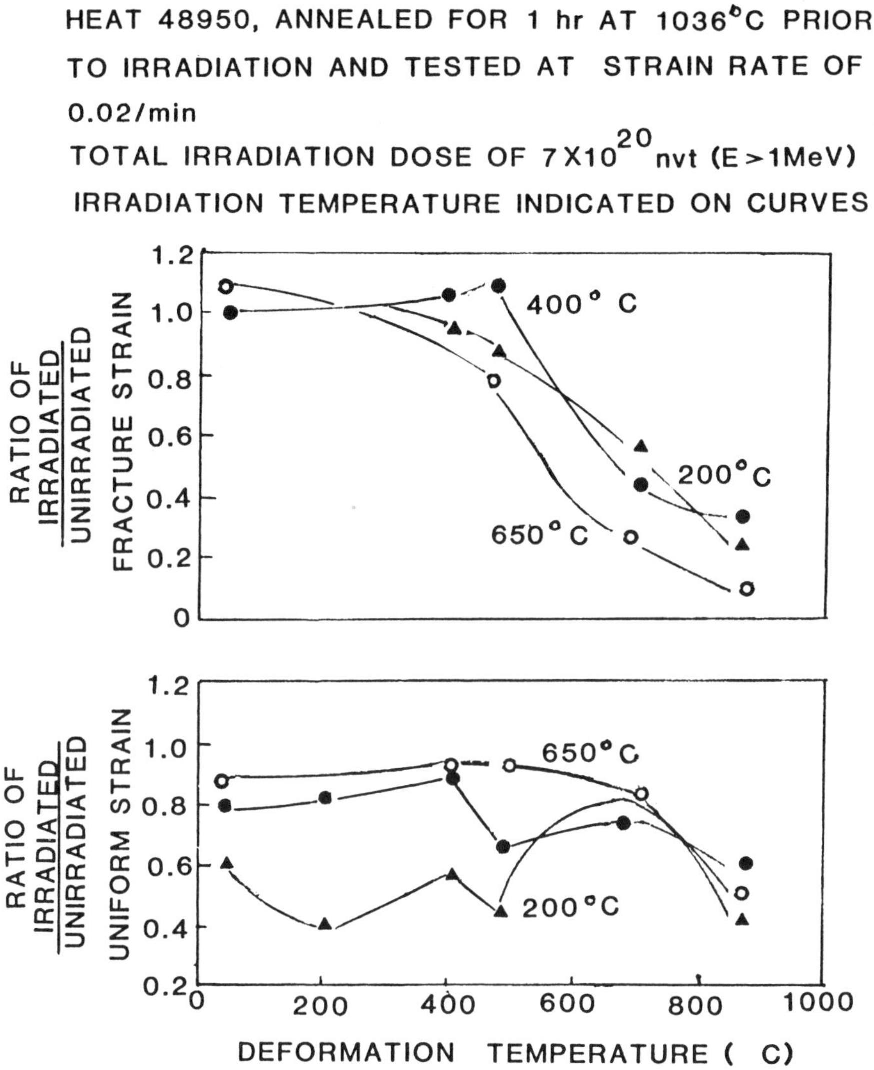

FIGURE 7.32 Effect of irradiation on the ductility of Type 304 stainless steel.

[From Martin, Ref. (7-40)]

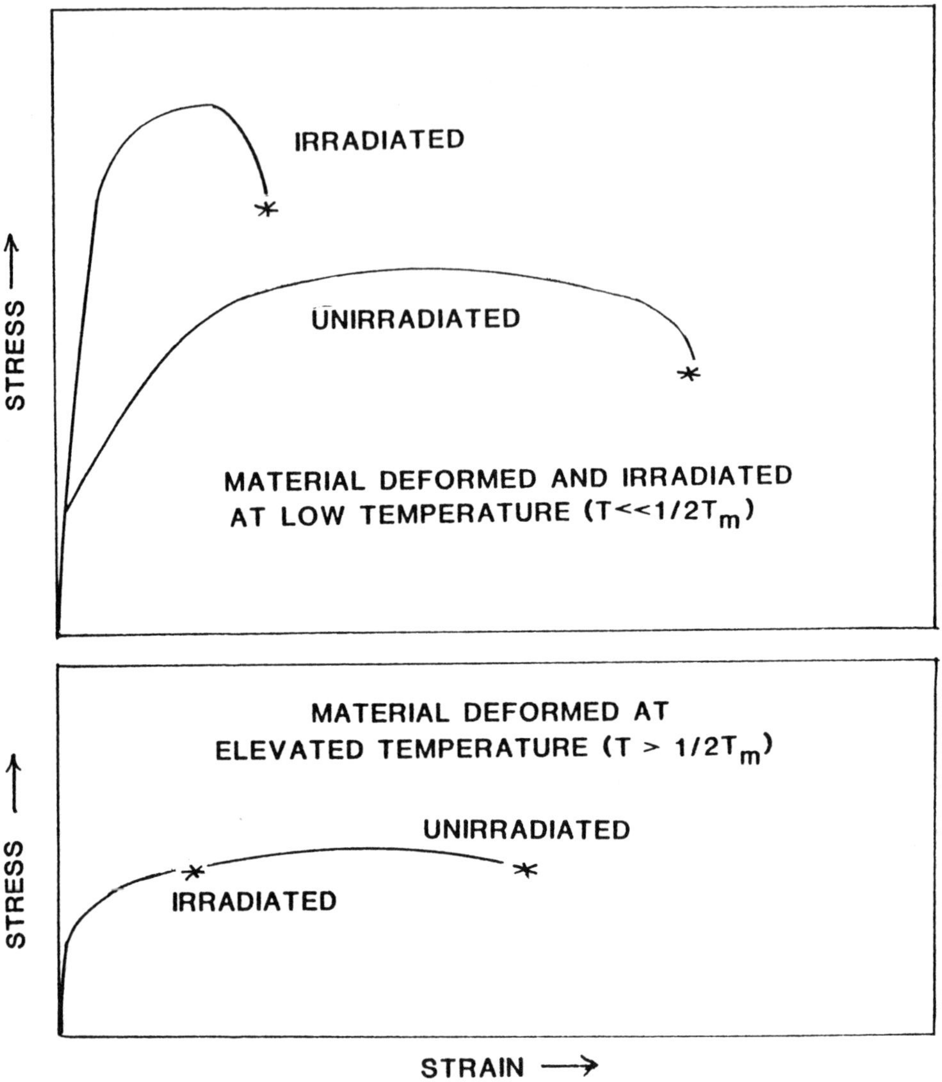

FIGURE 7.33 Effect of irradiation on the stress-strain curve.

[From Martin, Ref. (7-40)]

TABLE 7.14

Effect of Irradiation on the Mechanical Properties of Stainless Steels as a Function of Deformation and Irradiation Temperature

Mechanical Properties	Deformation Temperature		
	$T < 1/2\ T_m$ [a]		$T > 1/2\ T_m$
	Low Irradiation Temperature	High Irradiation Temperature	Low and High Irradiation Temperatures
Yield stress	large increase	no effect	no effect
Engineering ultimate stress	small increase	no effect	small decrease
True tensile stress	no effect	no effect	small decrease
True fracture stress	no effect	no effect	small decrease
True uniform strain	large decrease	no effect	decreased
True fracture strain	no effect	no effect	large decrease
Work hardening coefficient	large decrease	no effect	no effect

[a] T_m = absolute melting point of the alloy.

[From Martin, Ref. (7-40)]

The results of fatigue tests on Type 304 stainless steel at 430°C with strain amplitude selected to produce failure at 300-10^5 cycles are shown in Figure 7.34 [Reference (7-41)]. The data correlate to a relation of the form

$$\Delta\varepsilon_p = c\ N_f^{-m} \qquad (7.32)$$

where

c,m = material constants

The Coffin-Manson relation [Equation (7.24)] has been found to be valid for many metals and alloys at temperatures below the creep range [Reference (7-41)].

The effect of temperature on fatigue properties of Type 347 stainless steel have been described by Baldwin _et al_ [Reference (7-42)] and the data are plotted in Figures 7.35 through 7.37.

Kawasaki _et al_ [Reference (7-43)] studied the effect of temperature on fatigue strength, and the role of dislocations at high temperature. The fatigue strength takes a maximum value when the following conditions are satisfied

$$Dt = 4\ X^2/\pi^2 \qquad (7.33)$$

where

D = diffusion coefficient of solute atom = $D_0 \exp(-Q/RT)$

D_0 = constant which contains many factors that are temperature independent

t = time corresponding to N cycles

X = spacing (length) between cyclic stress

2X = mean spacing between parallel cyclic stress

Q = activation energy for diffusion

R = gas constant

T = absolute temperature

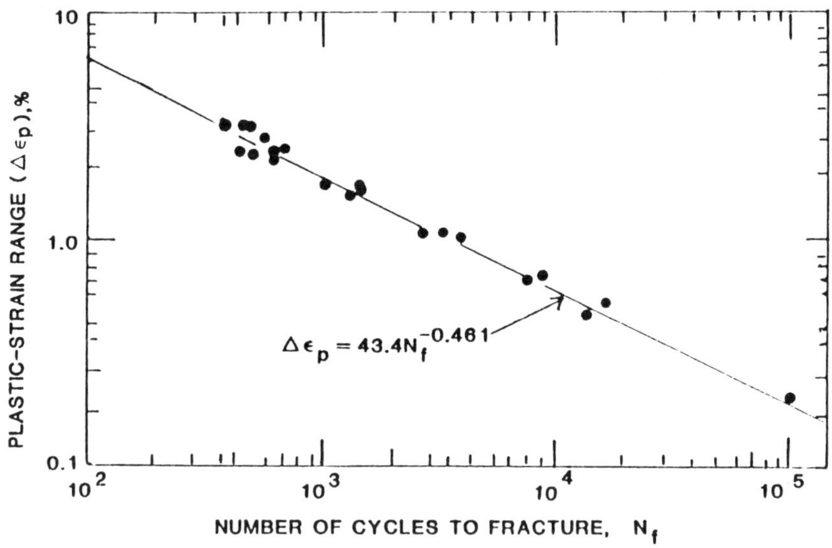

MATERIAL: ANNEALED 304 STAINLESS STEEL, HEAT NO. 55697

TEMPERATURE: 430°C

CONTROL MODE: TRIANGULAR WITH ZERO MEAN STRAIN

NOMINAL DIAMETRAL STRAIN RATE: 1.8×10^{-3} sec^{-1}

STARTING CONDITION: COMPRESSION FIRST

SPECIMEN: PUSH-PULL TYPE, HOURGLASS GAUGE SECTION, LONGITUDINAL SURFACE FINISH

FIGURE 7.34 Low-cycle fatigue-test results for 304 stainless steel at 430°C.

[Reproduced with permission from USAEC Report GEMP-63, "High Temperature Materials Program," General Electric Company, Cincinnati, Ohio, Dec. 30, 1966, Nucl. Mat. Prop. Of.]

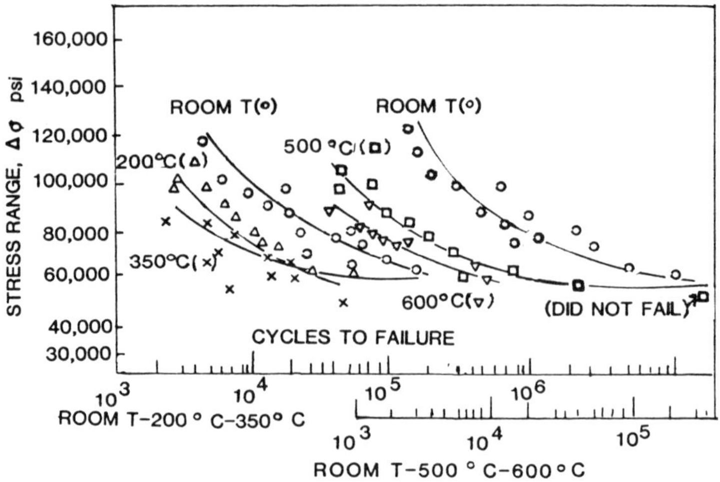

FIGURE 7.35 Stress range versus cycles to failure 347 stainless steel rod (Ht-11615).

[Reprinted with permission from the American Society for Testing and Materials, 1916 Race Street, Philadelphia, PA 19103"; Baldwin, Ref. (7-42)]

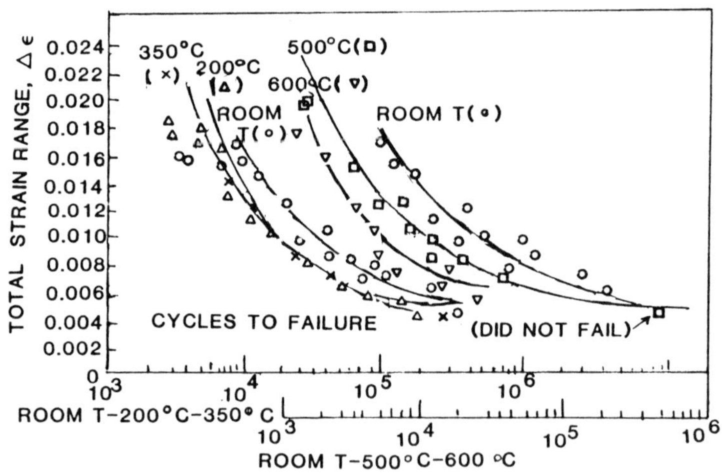

FIGURE 7.36 Total strain range versus cycles to failure 347 stainless steel rod (Ht-11615).

[From Baldwin, Ref. (7-42)]

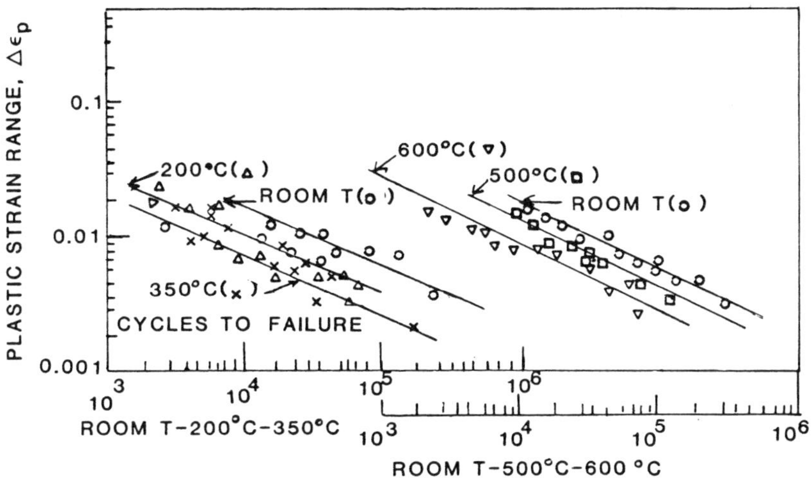

FIGURE 7.37 Plastic strain range versus cycles to failure 347 stainless steel rod (HT-11615).

[From Baldwin, Ref. (7-42)]

In summary, the effects of neutron irradiation on the fatigue resistance of reactor structural materials are not readily predicted from the response of tensile properties to irradiation. A possible increase in fatigue strength is suggested by the known increase in yield strength during irradiation. A reduction in fatigue life might be projected because of loss in ductility.

7.2 Radiation Effects on Light Water Reactor (LWR) Pressure Vessels

The reactor vessel for LWRs is typically made from rolled plate of a low alloy steel. For example the Wm. H. Zimmer BWR plant uses AA-533 Grade B steel (ASTM/ASME specifications). Although the inside walls of the reactor vessel are exposed to only a moderate fast neutron flux [10^9-10^{10} n/(cm^2)(sec)], damage can accumulate over the long service period of the vessel.

7.2.1 General effects of fast neutron collisions on steel. As discussed in Section 3.2 of Chapter 3, neutrons can produce Frenkel pairs in a metal. Both vacancies and interstitials can move as a result of thermal energy, and it has been concluded that interstitials and vacancies may run together and cancel each other, or they may condense on dislocations, causing embrittlement.

Near the original neutron hit, atoms collide energetically causing violent vibration of the lattice. Since lattice vibrational energy is a measure of temperature, there is a locally heated region termed a "thermal spike." Analysis of the thermal spike phenomenon is presented in detail in Chapter 8. The central portion of the spike is heated far above the melting point of the metal and the heat is dissipated into the surrounding metal in about 10^{-10} sec. When a small region of the lattice is rapidly heated in this way, thermal expansion will build up a tremendous pressure of around 2 x 10^5 atm. In steel, which has allotropic transformations that are influenced by the alloying elements present, it is not known what transformation characteristics will apply under these conditions--extremely rapid heating under extreme pressure followed by rapid cooling. Generally, it has been found that these conditions will increase the brittle to ductile transition temperature.

Since ferritic steels have a body-centered cubic crystal structure, the fracture mode changes from ductile to brittle with decreasing temperature. In a very narrow range of temperature, the impact strength of the body-centered cubic materials decreases rapidly and the temperature at which the impact strength has a specific value is generally known as the brittle to ductile transition temperature. Irradiation increases the transition temperature and, if it is increased to the operating temperature range, there is always a possibility of brittle failure.

Illustration 7.1 The fast neutron flux impinging upon the pressure vessel is 2.1×10^9 n/(cm^2)(sec) (>0.1 MeV). Calculate the fractional transmutation rate of iron atoms.

The transmutation rate per unit volume is

$$R = \Sigma_a \Phi$$
$$= N\sigma_a \Phi \quad \frac{\text{transmutations}}{(\text{cm}^3)(\text{sec})}$$

where

N = atomic density of iron

σ_a= microscopic absorption cross section which leads to transmutation

Thus the fractional change rate is

$$\frac{R}{N} = \frac{N\sigma_a\Phi}{N} = \sigma_a\Phi$$
$$= (2.62 \times 10^{-24}) \times (2.1 \times 10^9)$$
$$= 5.5 \times 10^{-15} \text{ sec}^{-1}$$
$$= 1.7 \times 10^{-7} \text{ year}^{-1}$$

which is too small to cause significant mechanical property changes.

7.2.2 Effects of neutron radiation on tensile and impact properties of steels. For irradiation temperatures to 500°F, Figure 7.38 shows that yield strength increases rapidly from neutron fluences of 1×10^{19} to 4×10^{19} n/cm^2 [Reference (7-44)]. For comparable conditions, the tensile strength increase varies from about 20,000 psi at 10^{19} n/cm^2 to 60,000 psi at 4×10^{19}n/cm^2 (>1MeV) [Reference (7-44)]. Changes in tensile properties of irradiated steels are listed in Table 7.15 [Reference (7-44)]. Changes in the stress-strain curves to irradiation can be seen in Figure 7.39 [Reference (7-45)]. Since the yield strength generally rises more than the tensile strength, the yield to tensile strength ratio is increased. The total elongation is reduced and the uniform elongation drops even more markedly. The marked loss of uniform elongation indicates a loss of work hardening ability on deformation. The nil-ductility transition (NDT) temperature fracture at which the metal changes from ductile to brittle fracture is increased by neutron irradiation as shown in Figure 7.40 for A212-B steels [Reference (7-46)]. Note that the steel is ductile at "high" temperature and brittle at "low" temperature, but that the temperature of transition increases with radiation.

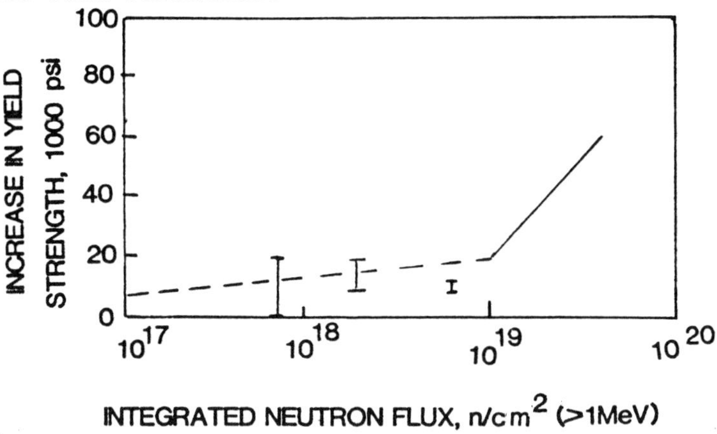

FIGURE 7.38 Yield strength increases of carbon and alloy steels after irradiation at temperatures to 500°F.

[Reproduced with permission from L.P. Trudean, Radiation Effects on Toughness of Ferritic Steels for Reactor Vessels, Rowman and Littlefield, Inc., New York, 1964]

TABLE 7.15

Tensile Properties of Irradiated Steels

Steel	Exposure, n/cm^2 (>1 MeV)	Irradiation Temp, °F	Yield Strength 10^3 psi	Tensile Strength, 10^3 psi	Uniform Elongation, %	Elongation %	Reduction of Area %	Reference
ASTM A106	0	--	40	76	18	--	--	7-48
(fine grain	2×10^{19}	580	81	102	8	--	--	7-48
0.24% C)	2×10^{19}	680	55	87	11	--	--	7-48
(deoxidized	2×10^{19}	760	48	82	12	--	--	7-48
with aluminum)	8×10^{19}	580	79	106	6	--	broke without necking down	7-48
	8×10^{19}	780	47	79	11	--	--	7-48
	0	--	47	74	--	34	--	7-49
	1×10^{20}	175	97	102	--	11.5	--	7-49
	0	--	58	75	19	--	67	7-50
	1×10^{19}	140	73	78	23	--	70	7-50
	5×10^{18}	570	62	78	20	--	69	7-50
ASTM A106	0	--	46	80	14	--	--	7-48
(coarse grain	2×10^{19}	580	93	115	8	--	--	7-48
0.24% C)	2×10^{19}	680	67	98	9	--	--	7-48
	2×10^{19}	760	43	84	14	--	--	7-48
	7×10^{19}	580	87	103	3	--	broke without necking down	7-48
	7×10^{19}	780	64	94	11	--	--	7-48
	0	--	48	83	--	29	--	7-49
	1×10^{20}	175	116	121	--	6.5	--	7-49
	0	--	50	85	20	--	67	7-50
	1×10^{19}	140	72	89	16	--	--	7-50
	2×10^{18}	570	54	86	18	--	68	7-50
ASTM A201	0	--	48	75	--	31	--	7-51
(0.17% C)	8×10^{18}	150	77	82	--	17	--	7-51
	1.5×10^{19}	150	78	83	--	15	--	7-51
	2×10^{19}	150	88	93	--	13	--	7-51
	4×10^{19}	150	94	98	--	9.5	--	7-51
	1×10^{19}	425-515	60	84	--	16	--	7-51
	2.5×10^{19}	525-700	54	83	--	21	--	7-51
ASTM A201A	0	--	40	67	--	49	48	7-52
	1.4×10^{20}	200-400	119	124	--	10	12	7-52
ASTM A201B	0	--	43	73	--	41	53	7-53
	4×10^{19}	600	46	76	--	29	50	7-53

Steel	Exposure, n/cm^2 (>1 MeV)	Irradiation Temp, °F	Yield Strength 10^3 psi	Tensile Strength, 10^3 psi	Uniform Elongation, %	Elongation %	Reduction of Area %	Reference
ASTM A212B	0	--	50	75	--	22	64	7-50
(0.2% C)	1×10^{19}	175	65	81	--	18	68	7-50
	1×10^{20}	175	94	97	--	5	26	7-50
	1×10^{19}	570	57	84	--	12	57	7-50
	0	--	40	75	25	--	--	7-54
	2×10^{19}	175	92	98	6	--	--	7-54
	2×10^{19}	560	76	102	9	--	--	7-54
	2×10^{19}	680	61	90	12	--	--	7-54
	2×10^{19}	760	56	84	14	--	--	7-54
	6×10^{19}	700	82	105	6	--	broke without necking down	7-54
	6×10^{19}	780	59	71	13	--	--	7-54
	1×10^{20}	175	109	116	4	--	--	7-54
E7016	0	--	59	73	16	--	--	7-54
weld metal	5×10^{18}	175	69	78	11	--	--	7-54
	5×10^{18}	600	61	77	17	--	--	7-54
	2×10^{19}	175	108	108	0	--	--	7-54
	6×10^{19}	700	83	94	12	--	--	7-54
	6×10^{19}	780	69	77	15	--	--	7-54
	1×10^{20}	175	115	115	0	--	--	7-54
En 2	0	--	38	--	--	--	70	7-55
(0.15% C)	1×10^{17}	175	44	--	--	--	65	7-55
	1.5×10^{18}	175	54	--	--	--	64	7-55
	6×10^{18}	175	68	--	--	--	45	7-55
	9×10^{18}	175	83	--	--	--	56	7-55
ASTM A302B	0	--	68	89	--	23	--	7-51
(1.4% Mn,	2.4×10^{19}	150	119	119	--	11	--	7-51
0.5% Mo)	3.6×10^{19}	150	122	123	--	10	--	7-51
	0	--	65	92	--	26	61	7-56
	4×10^{17}	500	72	95	--	24	60	7-56
	0	--	68	90	--	17	66	7-57
	7×10^{17}	500	77	102	--	16	61	7-57
	0	--	64	90	--	27	61	7-56
	4×10^{17}	700	68	93	--	26	61	7-56
E 9016	0	--	86	96	--	18	72	7-57
weld metal	6×10^{17}	--	88	101	--	19	67	7-57
A-301	0	--	41	66	23	--	--	7-54
(0.12% C, 1%	5×10^{18}	150	55	69	23	--	--	7-54
Cr, 0.5% Mo)	5×10^{18}	575	47	71	26	--	--	7-54
	1.5×10^{19}	740	53	78	18	--	--	7-54
	4.5×10^{19}	740	63	86	12	--	--	7-54
	4.5×10^{19}	700	91	103	--	--	--	7-54

Steel	Exposure, n/cm^2 (>1 MeV)	Irradiation Temp, °F	Yield Strength 10^3 psi	Tensile Strength, 10^3 psi	Uniform Elongation, %	Elongation %	Reduction of Area %	Reference
ASTM A203D	0	--	67	86	--	25	--	7-51
(3.5% Ni)	2.4×10^{19}	150	122	123	--	10	--	7-51
	3.6×10^{19}	150	133	133	--	9	--	7-51
ASTM A353	0	--	92	119	22	30	--	7-49/54
(9% Ni)	1.7×10^{19}	175	138	148	11	18	--	7-49/54
	1×10^{20}	175	183	184	2	7	--	7-49/54
	0	--	114	138	--	17	--	7-51
	2.4×10^{19}	150	156	161	--	11	--	7-51
	2.8×10^{19}	150	167	172	--	11	--	7-51
HY-65	0	--	95	112	--	20	--	7-51/58
(Ni-Mo-Cu-V)	1.8×10^{19}	150	148	148	--	10	--	7-51
	3.4×10^{19}	150	142	142	--	9	--	7-58
HY-65	0	--	128	138	--	17	--	7-58
(austempered)	3.4×10^{19}	150	165	166	--	7	--	7-58
HY-65	0	--	146	150	--	16	--	7-58
(quenched and tempered)	3.4×10^{19}	150	198	198	--	8	--	7-58
HY-65	0	--	91	106	--	16	68	7-57
	8×10^{17}	500	113	131	--	15	59	7-57
E 10016	0	--	102	108	--	15	64	7-57
weld metal	7×10^{17}	--	100	116	--	14	61	7-57
HY-80	0	--	103	117	--	17	--	7-51
(Ni-Cr-Mo)	2.5×10^{19}	150	156	161	--	9	--	7-51
	5×10^{19}	150	176	176	--	8	--	7-51
Smithweld 91	0	--	124	133	--	14	--	7-51
	2.4×10^{19}	150	187	187	--	8	-	7-51
	4.5×10^{19}	150	203	206	--	7	--	7-51
U.S.S. T-1	0	--	120	129	8	14	--	7-49/54
	1.7×10^{19}	175	170	171	0.5	4.5	--	7-49/54
	1×10^{20}	175	186	187	0.3	4	--	7-49/58
	0	--	100	116	--	20	--	7-59
	1.3×10^{21}	100	180	180	--	5	--	7-59
	0	--	129	140	--	15	--	7-51
	2.3×10^{19}	150	184	186	--	9	--	7-51
	3.3×10^{19}	150	187	189	--	9	--	7-51
A.O. Smith 1146	0	--	73	108	--	25	--	7-51
(0.2% C, 1.2%	1.3×10^{19}	--	136	139	--	10	--	7-51
Mn, 0.5% Ni, 0.2% V)	2.8×10^{19}	--	140	144	--	10	--	7-51

[From Trudean, Ref. (7-44)]

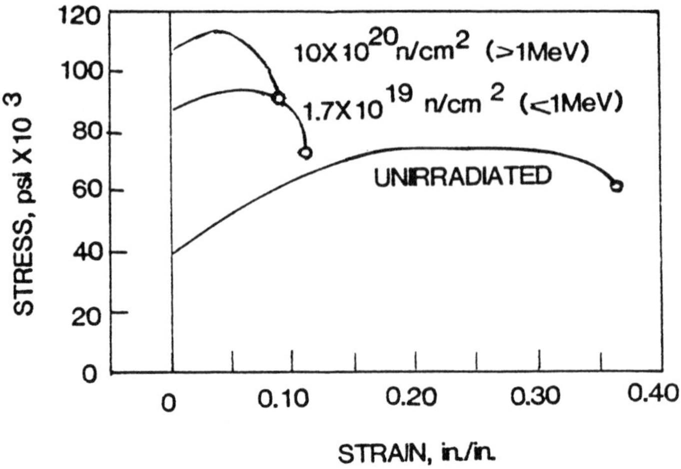

FIGURE 7.39 Load-elongation curves of irradiated (at less than 200°F) ASTM A212B carbon-silicon steel show that uniform elongation decreases with increasing neutron exposure. The drop in load at the yield point disappears for higher neutron exposures, but the general form of the curve is somewhat similar to the one for unirradiated steel.

[Taken from R.G. Berggren, "Radiation Effects in Ferritic Steels," USAEC Conf. on the Status of Radiation Effects Research on Structural Materials and Implications to Reactor Design, USAEC Report TID-7588, 1960]

7.2.3 Effect of increased irradiation temperature. With progressively higher irradiation temperatures, the shift in transition temperature at equivalent fast fluence is continuously lower as shown in Figure 7.41 [Reference (7-47)]. The NDT shift progressively decreases with increasing irradiation temperature because the less stable defect clusters anneal out at the higher temperature; only the most stable defect clusters remain and consequently only minor changes in the transition temperature can be expected. From Figure 7.41 it is clear that the steel has to be irradiated at temperatures in the 1200°F to 1400°F range to avoid any increase in NDT temperature.

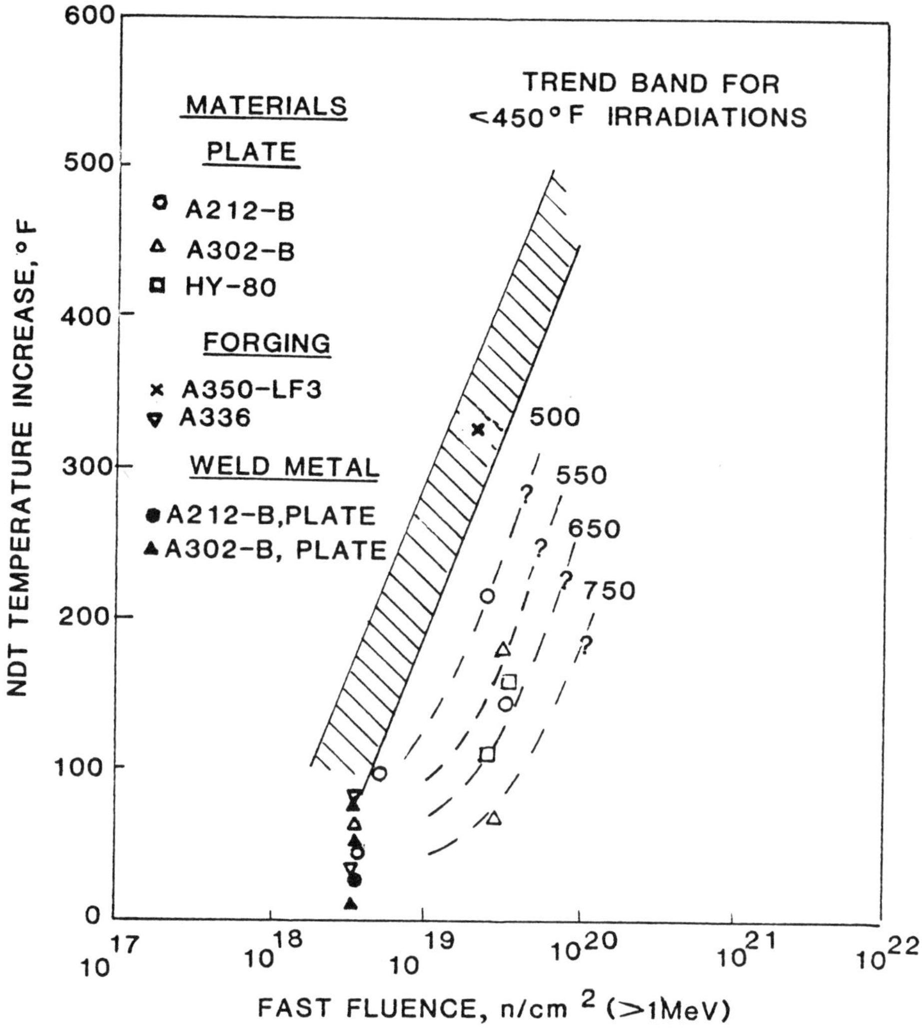

FIGURE 7.40 Increase in the NDT temperatures of steels resulting from irradiation at temperatures above 450°F.

[Reproduced with permission from L.E. Steele and J.R. Hawthorne, "Neutron Embrittlement of Reactor Pressure Vessel Steels," Ch. 16, Materials and Fuels for High Temperature Nuclear Energy Applications, The MIT Press, Cambridge, MA 1964]

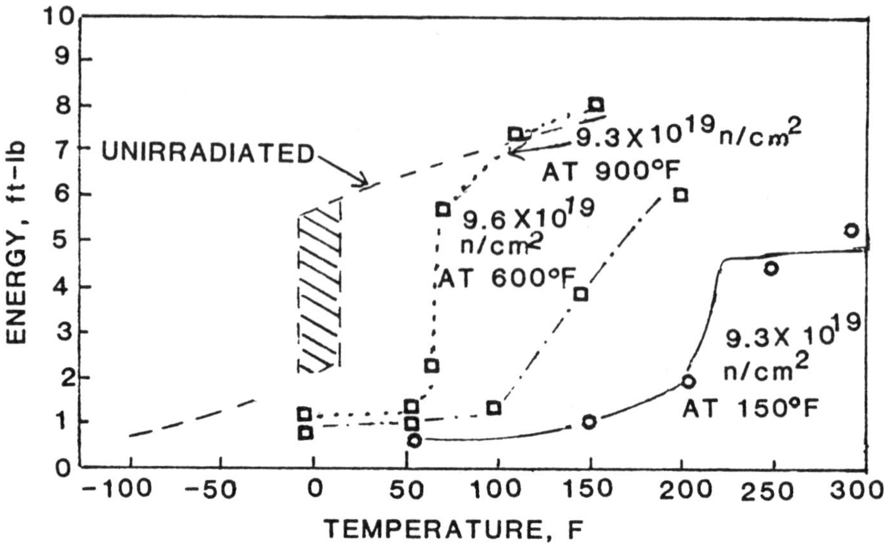

(a) 2.25 CR-1 MO STEEL

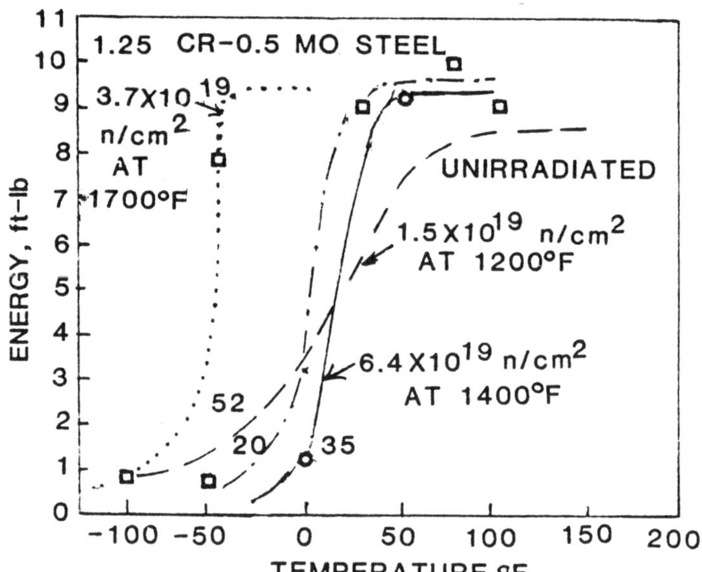

(b) 1.25 CR-0.5 MO STEEL

FIGURE 7.41 Effect of neutron radiation and irradiation temperature on the impact properties of chromium-molybdenum steels.

["Reprinted with permission from the American Society for Testing and Materials, 1916 Race Street, Philadelphia, PA 19103"; Lowe, Ref. (7-47)]

The combination of gamma heating and neutron heating of the wall of the reactor vessel can create appreciable thermal stresses. The alternate heating and cooling of the pressure wall will give rise to cyclic stresses.

<u>7.2.4 Control of radiation damage</u>. Since irradiation embrittlement increases the transition temperature, the concern is that the NDT temperature may increase to the operating temperature of the pressure vessel, therefore making it brittle and susceptible to fracture. It has been found that some of the damage can be mended by annealing the reactor vessel in place. The annealing temperature will vary from one reactor design to another. The maximum limit ranges between 600°F to 750°F. Although such practice is not yet common in the power reactor industry, some previous work has been done along this line [Reference (7-44)]. Most of the annealing procedures involve removing the core of the reactor and also, if possible, the thermal shield. Then heated inert gas of nitrogen is blown through the vessel. It has been found that annealing after irradiation below 500°F is more effective than the same anneal after irradiation at a higher temperature. It is a standard practice to limit the minimum operating temperature of pressure vessels, while under pressure, to the temperature corresponding to NDT + 60°F [Reference (7-60)].

Also through control of composition, especially that of the residual elements copper and phosphorus, it is possible to produce steels having consistently high irradiation embrittlement resistance. Copper contents in the range of .08% to .14% generally indicate steels of relatively low sensitivity while levels greater than .19% or .20% are associated with high sensitivity. Superior radiation resistance has been demonstrated in a controlled composition of A533 steel containing .03% copper and .005% of phosphorus [Reference (7-61)].

Illustration 7.2 For illustrative purpose, take the temperature and fast neutron flux at the inside wall of pressure vessel to be 500 °F and 2.1×10^9 n/(cm^2)(sec). By using Figure 7.40, estimate a) the total fluence and NDT temperature increase for a reactor of 80% load factor and 40 years lifetime; b) the flux required to make the pressure vessel susceptible to radiation damage.

a. Total fluence $= 2.1 \times 10^9 \times 40 \times 365 \times 24 \times 3600 \times 0.8$
$= 2.12 \times 10^{18}$ n/cm^2

From Figure 7.40, NDT temperature increase ≅ 50 °F
Normally the NDT temperature is in the range 0~50 °F, depending on the kind of steel. Taking the unirradiated NDT to be 50 °F,

NDT + 60 °F = (50 °F + 50 °F) + 60 °F = 160 °F

which is below operating temperature.

b. To make NDT temperature increase become significant, we have

NDT + 60 °F = 500 °F

Hence the required NDT after irradiation is 440 °F. The NDT temperature increase is thus about 390 °F. From Figure 7.40, the fluence required is 10^{20} n/cm^2. Thus the required fast neutron flux

$$= \frac{10^{20}}{40 \times 365 \times 24 \times 3600} \cong 10^{11} \text{ n/(cm}^2\text{)(sec)}$$

which is two orders of magnitude higher than the flux in part a.

REFERENCES

(7-1) Brickner, K.G., "Stainless Steels for Room and Cryogenic Temperatures," in Selection for Stainless Steels, American Society for Metals, Metals Park, Ohio, 1968.

(7-2) Van Vlack, L.H., Materials Science for Engineers, Addison-Wesley, Reading, Massachussetts, Ch. 1, 1970.

(7-3) Olander, D.R., Fundamental Aspect of Nuclear Reactor Fuel Elements, Energy Research and Development, Report TID-26711-P1, Ch. 18, 1976.

(7-4) Hertzberg, R., Deformation and Fracture Mechanics of Engineering Materials, John Wiley and Sons, New York, Ch. 1, 7, and 12, 1976.

(7-5) Holloman, J.H., Tensile Deformation, Trans. AIME, Vol. 162, 1945, pp. 268-290.

(7-6) Steel Products Manual, Section 24, "Stainless Steel and Heat Resisting Steel," American Iron Steel Institute, December 1974.

(7-7) Dieter, G.E., Mechanical Metallurgy, McGraw-Hill Book Company, New York, Ch. 9, 1961.

(7-8) Hoke, J.H., "Mechanical Properties of Stainless Steels at Elevated Temperature," in Handbook of Stainless Steel by D. Peckner and I.M. Bernstein, McGraw-Hill Book Company, New York, Ch. 21, 1977.

(7-9) Spretnak, J.W., Fontana, M.G., and Brooks, H.E., "Notched and Unnotched Tensile and Fatigue Properties of Ten Engineering Alloys," Trans. Amer. Soc. Metals, 43, 1951, p. 547.

(7-10) a. Smith, D.L., et al, "In Symposium on Structural Materials for Service at Elevated Temperatures in Nuclear Power Generation," ASME Winter Meeting, Nov. 30-Dec. 3, 1975, Houston, Texas, pp. 290-301.

b. Brinkman, C.R., et al, "Estimates of Creep-Fatigue Interaction in Irradiated and Unirradiated Austenitic Stainless Steel," Nucl. Tech., Vol. 16, 1972, pp. 297-307.

(7-11) Knoepfel, H., Eric-Trapani (Sicily), "Tokamak Reactors for Breakeven," Pergamon Press, Sept. 21-Oct. 1, 1976, pp. 449-484.

(7-12) Fish, R.L., and Holmes, J.J., "Tensile Properties of Annealed Type 316 Stainless Steel After EBR-II Irradiation," J. Nucl. Materials, 46, 1973, pp. 113-120.

(7-13) Packan, Nicolas H., and Farrell, Kenneth, "Radiation-Induced Swelling in an Austenitic Alloy: Observations and Interpretation of the Effects of Helium," Nuclear Technology/Fusion, 3, May 1983, pp. 392-404.

(7-14) Das, S.K., "The Role of Inert Gases in First Wall Phenomena in Fusion Devices," Proc. Harwell Consultants Symp. Inert Gases in Metals and Ionic Solids, AERE-R9733, Vol. II, p. 409, S.F. Pugh, Ed., U.K. Atomic Energy Research Establishments, Mar. 1980.

(7-15) Wohler, A., "Versuche uber die Festtigkeit der Eisenbahnwagen-Achsen," Zeitschrift fur Bauwesen, Vol. 10, 1860.

(7-16) Bloom, E.E., "Mechanical Properties of Materials in Fusion Reactor First-Wall and Blanket Systems," J. Nucl. Mater., 85 and 86, 1979, 795.

(7-17) Farrell, K., "Experimental Effects of Helium on Cavity Formation During Irradiation - A Review," Rad. Eff., 53, 1980, 175.

(7-18) Palmgren, A., Bertschrift des Vereines Ingenieure, 58, 1924, p. 339.

(7-19) Miner, M.A., "Cumulative Damage in Fatigue," Jour. Appl. Mechanics, Vol. 12, P.A-159, 1954.

(7-20) Sandor, B.I., Fundamentals of Cyclic Stress and Strain, University of Wisconsin Press, Madison, Wisc., 1972.

(7-21) Smith, R.W., Hirschberg, M.H., and Manson, S.S., "Fatigue Behavior of Materials Under Strain Cycling in Low and Intermediate Life Range," NASA, TN-D-1574, April 1963.

(7-22) Manson, S.S., and Hirschberg, M.H., in "Fatigue: An Interdisciplinary Approach," Syracuse Univ. Press, New York, 1964, p. 133, proceedings 10th Sagamore Conference.

(7-23) Morrow, J.O. Dean, "In Internal Friction, Damping and Cyclic Plasticity," ASTM STP 378, 1965, p. 45, American Society for Testing and Materials.

(7-24) Coffin, L.F., Jr., "A Study of the Effects of Cyclic Thermal Stresses on Ductile Metals," Trans. Amer. Soc. Mechanical Engrs., Vol. 76, 1954, p. 931-950.

(7-25) Tavernelli, J.F., and Coffin, L.F., Jr., "A Compilation and Interpretation of Cyclic Strain Fatigue Tests on Metals," Trans. ASM, Vol. 51, 1959, p. 438.

(7-26) Landgraf, R.W., "In Achievement of High Fatigue Resistance in Metals and Alloys," ASTM STP-467, 1970, p. 3, American Society for Testing and Material.

(7-27) Manson, S.S., Thermal Stress and Low Cycle Fatigue, McGraw-Hill Book Company, New York, 1966, p. 270.

(7-28) Natesan, K., et al, "In Symposium on Structural Materials for Service at Elevated Temperatures in Nuclear Power Generation," ASME Winter Meeting Nov. 30-Dec. 3, 1975, Houston, Texas, pp. 302-315.

(7-29) Halford, G.R., and Manson, S.S., "Applications of a Method of Estimating High Temperature Low Cycle Fatigue Behavior," NASA TMX-52357, Oct. 1967, Lewis Research Center, Cleveland, Ohio.

(7-30) Conway, J.B., "Evaluation of Plastic Fatigue Properties of Heat-Resistant Alloys," GEMP-740, Dec. 1969, General Elec. Company, Cincinnati, Ohio, Nucl. System Program.

(7-31) Halford, G.R., "Private Communication," NASA Lewis Research Center, National Aeronautic and Space Admin., Cleveland, Ohio, Nov. 1971.

(7-32) Brinkman, C.R., Korth, G.E., and Beeston, J.M., "In Effects of Radiation of Substructure and Mechanical Properties of Metals and Alloys," ASTM STP 529, American Society for Testing and Materials, 1973, pp. 473-492.

(7-33) Beeston, J.M., and Brinkman, C.R., "In Irradiation Effects on Structural Alloys for Nuclear Reactor Applications," ASTM STP 484, American Society for Testing and Materials, 1970, pp. 419-450.

(7-34) Kangilaski, M., "Radiation Effects in Structural Materials," Reactor Materials, Vol. 13, No. 1, Quarterly Technical Progress Review prepared for Division of Technical Information, USAEC, by Battelle Memorial Institute, Columbus Lab., 1970.

(7-35) Keller, D.L. (comp), "Progress on Development of Materials and Technology for Advanced Reactors During January through March 1969," USAEC Report BMI-1862, Battelle-Columbus, April 1969.

(7-36) Conway, J.B., "Evaluation of the Plastic Fatigue Properties of Heat-Resistant Alloys," USAEC Report GEMP-1010, Dec. 1968, pp. 161-198, General Electric Company, Cincinnati, Ohio, Missile and Space Division.

(7-37) Berling, J.T., and Conway, J.B., "A Proposed Method for Predicting the Low-Cycle Fatigue Behavior of 304 and 316 Stainless Steel," Trans. Met. Soc. AIME, May 1969, pp. 1137-1140.

(7-38) Bement, A.L., et al, "Irradiation Damage to Reactor Metals," Reactor Fuel and Materials Branch of USAEC Division of Reactor Development and Technology, Quarterly Progress Report 1968, USAEC Report BNWL-919, Nov. 1968, Battelle-Northwest; Richland, Wash. Pacific Northwest Lab.

(7-39) Reactor Development Program Progress Report, July 1968, USAEC Report ANL-7478, Argonne National Laboratory, August 28, 1968, pp. 56-57.

(7-40) Martin, W.R., and Weir, J.R., "In Flow and Fracture of Metals and Alloys in Nuclear Environments," ASTM STP 380, American Society for Testing and Materials, 1965, pp. 251-268.

(7-41) "High Temperature Materials Program," USAEC Report GEMP-63, General Electric Company, Cincinnati, Ohio Dec. 30, 1966, Nucl. Mat. Prop. Of.

(7-42) Baldwin, E.E., Sokol, G.J., and Coffin, L.F., Jr., "Cyclic Strain Fatigue Studies on AISI Type 347 Stainless Steel," ASTM proceedings, Vol. 57, 1957, pp. 567-581.

(7-43) Kawasaki, T., Izumi, H., and Sawaki, Y., "A Theoretical Explanation for the Increase in Fatigue Strength of Mild Steel at Elevated Temperature," Trans. ASM, Vol. 60, 1967, p. 707.

(7-44) Trudean, L.P., Radiation Effects on Toughness of Ferritic Steels for Reactor Vessels, Rowman and Littlefield, Inc., New York, 1964.

(7-45) Berggren, R.G., "Radiation Effects in Ferritic Steels," USAEC Conference on the Status of Radiation Effects Research on Structural Materials and Implications to Reactor Design, USAEC Report TID-7588, 1960.

(7-46) Steele, L.E., and Hawthorne, J.R., "Neutron Embrittlement of Reactor Pressure Vessel Steels," Ch. 16, Materials and Fuels for High Temperature Nuclear Energy Applications, The MIT Press, Cambridge, MA, 1964.

(7-47) Lowe, A.L., Jr., "Effects of Radiation on Two Low-Alloy Steels at Elevated Temperatures," ASTM STP No. 34, 1963.

(7-48) Wilson, J.C., Berggren, R.G., Grizzel, F.M., and Humphries, J.T., "HRP Radiation Metallurgy, Notch Impact Properties," USAEC Report ORNL-2615, 1958.

(7-49) Berggren, R.G., and Wilson, J.C., "Recent Data on the Effects of Neutron Irradiation on Structural Metals and Alloys," Report ORNL-CF-56-11-1, 1957.

(7-50) Wilson, J.C., and Berggren, R.G., "Effects of Neutron Irradiation in Steels," Proc. ASTM, 55, 1955, 689.

(7-51) Trudeau, L.P., "Irradiation on Some Pressure-Vessel Steels," ASTM STP No. 276, 1960.

(7-52) Mehan, R.L., and Baldwin, E.E., "Effect of Neutron Irradiation on Notched Bend and Tensile Properties of ASTM 201 A Carbon Steel," USAEC Report KAPL-1874 1957.

(7-53) Bartz, M.H., "Performance of Metals During Six Years Service in the Materials Testing Reactor," Second United Nations Conference on the Peaceful Uses of Atomic Energy, Geneva, Switzerland, 5, 1958, 466.

(7-54) Wilson, J.C., "Effects of Irradiation on the Structural Materials in Nuclear Power Reactors," Second United Nations Conference on the Peaceful Uses of Atomic Energy, Geneva, Switzerland, 5, 1958, 431.

(7-55) Churchman, A.T., Hull, D., and Mogford, I.L., "The Tansik Properties of Steel Irradiated with Neutrons," Atomic Energy Research Establishments, Harwell, England, Report AERE M/R 2515, 1958.

(7-56) Baldwin, E.E., "Neutron Radiation Effects on Tensile and Impact Properties of ASTM A302B Steel," Mechanical Engineering, 79, 1957, 261.

(7-57) Alger, J.V., and Srupien, L.M., "Neutron Radiation Embrittlement at 500°F and 650°F of Reactor Pressure Steels," ASTM STP No. 276, 1960.

(7-58) Trudeau, L.P., "Effects of Neutron Irradiation on Mechanical Properties of Ferritic Steels and Irons, Second United Nations Conference on the Peaceful Uses of Atomic Energy, Geneva, Switzerland, 5, 1958, 475.

(7-59) Balai, N., Kettles, T.E., and Bailey, R.E., "Reactor Pressure Vessel Design for Nuclear Applications," ASTM STP No. 233, 1958.

(7-60) Weisman, J., Elements of Nuclear Reactor Design, Elsevier, New York, 1977.

(7-61) Steele, L.E., "Critical Aspects of Neutron Irradiation Embrittlement of Pressure Vessel Steels and Weldments," Nucl. Met., 16, 1970, pp. 270-291.

PROBLEMS FOR CHAPTER 7

1. A 10-cm-long rod is loaded with 10^3 Newton force. The diameter decreases by 90% of its original diameter. If the original diameter is 2 mm, calculate the following:

   a. The final length of the rod.

   b. The true stress and true strain at this load.

   c. The engineering stress and strain at this load.

2. True stress is defined as the load divided by the instantaneous area. Verify that $\sigma_{true} =$

   $\sigma_{eng}(1 + \varepsilon_{eng})$.

3. Show that the true strain is equal to the natural logarithm of one plus the engineering strain, i.e.

   $\varepsilon_{true} = \ln(1 + \varepsilon_{eng})$.

4. Prove that the true plastic strain at necking instability is numerically equal to the strain hardening coefficient, i.e.

   $\varepsilon_{true} = \bar{n}$

   (Hint $dP = \sigma_{true} dA + A d\sigma_{true} = 0$ at the necking instability.)

5. What is the total strain range that a bar would be able to tolerate for a cyclic strain specimen before failing after 10^3 cycles of $E = 200 \times 10^9$ N/m^2

σ_f (monotonic fracture strength) = 2×10^9 N/m^2
ε_f (monotonic true fracture ductility) = 0.7
$\bar{n}'$ (cyclic) = 0.15

6. Turbine blades are suddenly heated from 600°C to 800°C. If the blades are under a stress of 3×10^8 N/m^2 and the thermal expansion coefficient is 1×10^{-5}/°C, the modulus of elasticity 2×10^{11}, and the fracture strain is 0.25, determine

a. The elastic strain.

b. The total strain.

c. The plastic strain.

d. Number of cycles the engine will make before failure if the plastic strain is given by

$$\varepsilon_p = \frac{\varepsilon_f}{2} (N)^{-0.5}$$

7. Use the flux values in part a. and b. of Illustration 7.2, calculate the equilibrium fraction of displaced atoms in pressure vessel to see the range of fractional change in which properties of iron are susceptible to radiation damage. (Use recombination coefficient $A = 2.62 \times 10^{-28}$ cm^3/sec.)

8. Assume a fast neutron flux in the pressure vessel wall of 5×10^9 n/(cm^2)(sec). At the end of 40 years of operation, at what minimum temperature should the pressure vessel be maintained in order to avoid brittle fracture while subjected to high pressure? (Load factor of the reactor = 75%)

8 NUCLEAR FISSION EFFECTS

This chapter deals directly with phenomena related to the nuclear fission process. After a discussion of the fission process, the events of thermal spikes and nuclear fuel swelling are presented.

8.1 The Fission Process

The fission process is basic, of course, to many areas of nuclear physics and engineering. While a detailed understanding of the theory of fission is not necessary to an engineering approach to a study of radiation effects, the results of the fission process certainly are. This section of the chapter presents the details of the fission process which are required for the later studies.

8.1.1 What causes fission.* Although not entirely essential for engineering purposes, the mechanisms at play which allow the fission process to occur add an element of understanding as to why certain nuclides, such as uranium-235, readily fission with low energy (thermal) neutrons, while its sister nuclides, such as uranium-238, do not.

---

* Most of this discussion is taken from Reference (8-1).

The binding energy (B.E.) is the energy needed to separate the nucleons of a given nucleus from each other. It is found that the measured atomic mass of any isotope is less than the sum of the masses of its constituent particles in the free state. As an example, consider the $^{4}_{2}He$ nucleus, which contains two protons and two neutrons.

proton mass = 1.007277 amu = m_p
neutron mass = 1.008665 amu = m_n
electron mass = 0.0005488 amu = m_e
(1 amu = 1 atomic mass unit = 1.660×10^{-24} gm)

The sum of the helium nucleus constituents is (1.007277 + 1.008665) x 2 = 4.03188 amu. The experimental mass of the atom is 4.00387 amu which, when corrected by 2 x 0.00055 amu for the orbital electrons gives 4.00277 amu. The mass defect is Δm = 4.00277 - 4.03188 = -0.02911 amu. The binding energy is the energy equivalence of this mass as calculated by Einstein's formula

$$B.E. = \Delta m\ c^2 \tag{8.1}$$

The energy equivalence conversion factor is 931.478 MeV/amu where 1 MeV = 1.602×10^{-13} joules. Thus, in this example with the helium nucleus, the binding energy is (0.02911) x (931.478) = 27.1 MeV. The average binding energy per nucleon is just

$$B.E.\ \text{per nucleon} = \frac{\Delta m\ c^2}{A} \tag{8.2}$$

which in this example is $\frac{27.1}{4}$ = 6.78 MeV/nucleon. The binding energy can thus be calculated from the mass data as

$$B.E. = \Delta m\ c^2 = c^2[Z(m_p + m_e) + (A-Z)m_n] - M(A,Z)c^2 \tag{8.3}$$

where Z is the atomic number, A is the mass number, and M(A,Z) is the experimentally determined atomic mass. A curve of the average binding energy per nucleon versus mass number A can thus be constructed from the measured masses, and is drawn in Figure 8.1. Note that

$$\frac{B.E.}{A} \sim 8 \text{ MeV/nucleon for large A} \tag{8.4}$$

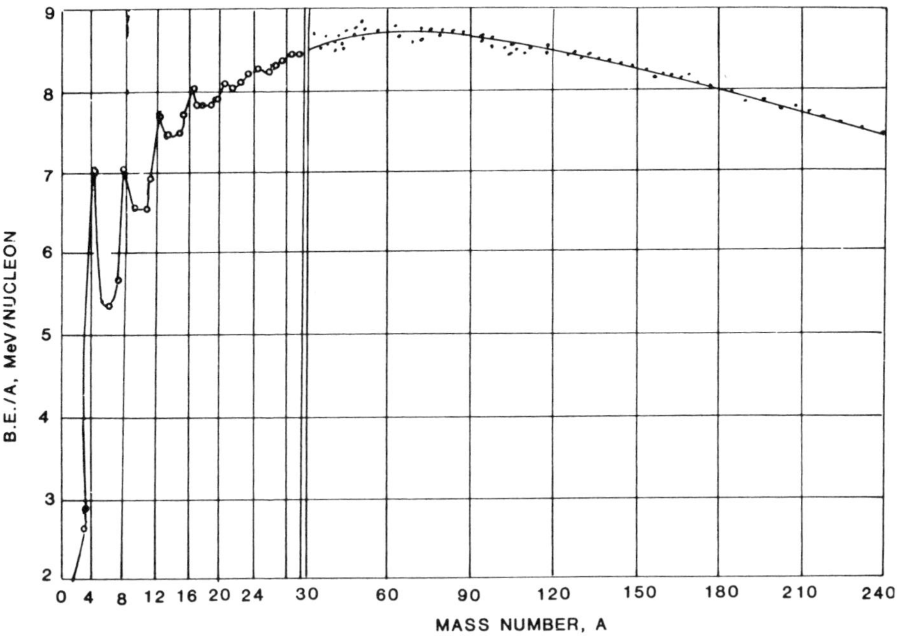

FIGURE 8.1 Average binding energy B.E./A in MeV per nucleon for the naturally occurring nuclides (and ^{8}Be), as a function of mass number A.

[Reproduced with permission from R.D. Evans, The Atomic Nucleus, McGraw-Hill Book Company, Inc., 1955, Chapter 11]

To develop a semiempirical mass formula, by what is known as the liquid-drop model of the nucleus [Reference (8-1)], consider the binding energy to be made up of a number of terms

$$B.E. = B_0 + B_1 + B_2 + \ldots \tag{8.5}$$

Each term in the series represents some characteristics of

nuclei. In the liquid-drop model, assume the following:

1. The nucleus is like a droplet of incompressible matter, and all nuclei have the same density.

2. Neglect spin and charge on forces between nucleons.

3. Nuclear forces have short range character and are effective only between nearest neighbors; each nucleon interacts with all its nearest neighbors.

The first order term is identified as a "saturated exchange force," or simply the "volume" energy. Since $A = \rho V$, where ρ is the density of nuclear matter and V is the volume, a term proportional to the volume is

$$B_0 = a_v A \tag{8.6}$$

where a_v is a coefficient to be determined. In other words, as a first order term, take B.E./A = constant.

Next consider surface effects. Nucleons on the surface have fewer near neighbors than those deep inside the nucleus, therefore, a deficit in binding energy exists for these surface nucleons. Now

$$A = \rho V = \frac{4}{3} \pi R^3 \rho \tag{8.7}$$

where R = radius of the (assumed) spherical nucleus so that

$$R = R_0 A^{1/3} \tag{8.8}$$

under the assumption of constant density ρ, where R_0 is a constant determined from experiments, usually taken as 1.3×10^{-15}m. The surface area of the nucleus is

$$4\pi R^2 = 4\pi R_0^2 A^{2/3} \tag{8.9}$$

Then the number of nucleons on the surface would be approximately

$$\frac{4\pi R^2}{\pi R_0^2} = \frac{4\pi R_0^2 A^{2/3}}{\pi R_0^2} = 4A^{2/3} \qquad (8.10)$$

We, therefore, introduce a negative correction term B_1 representing the loss of binding energy by the nucleons at the surface

$$B_1 = -a_s A^{2/3} \qquad (8.11)$$

where a_s is a coefficient to be determined.

The only known long-range force in nuclei is the Coulomb repulsion between protons. Regard the total nuclear charge (Ze) as spread approximately uniformly throughout the nuclear volume, then the charge density ρ' is

$$\rho' = \frac{Ze}{\frac{4}{3}\pi R^3} = \text{constant} \qquad (8.12)$$

Coulomb's Law for charge q_1, q_2 separated by a center-to-center distance of r is

$$F = \frac{q_1 q_2}{4\pi\varepsilon_0 r^2} \qquad (8.13)$$

where ε_0 is the permittivity of free space.

Consider the process of buildup of charge by bringing in increments of charge in the form of spherical shells from infinity as indicated in Figure 8.2. The charge in a sphere of radius r' is $z = \rho' \frac{4}{3}\pi r'^3$. The work done in adding a spherical mass element with charge $\Delta z = \rho'\Delta V$ is thus

$$W = \int_\infty^{r'} \underline{F} \cdot d\underline{r} = \int_\infty^{r'} \frac{z\Delta z}{4\pi\varepsilon_0 r^2}\, dr = -\frac{z\Delta z}{4\pi\varepsilon_0 r'}, = (\rho' \frac{4}{3}\pi r'^3) \times$$
$$(\rho'\pi 4r'^2 dr')(\frac{-1}{r'})(\frac{1}{4\pi\varepsilon_0}) = -\frac{3}{4\pi\varepsilon_0} (\frac{4\pi\rho'}{3})^2 r'^4 dr' \qquad (8.14)$$

Hence the total work done against Coulomb forces in creating a homogeneous spherical charge distribution is

$$W = \int_0^R \Delta W = -\frac{3}{4\pi\varepsilon_0}\left(\frac{4\pi\rho'}{3}\right)^2 \int_0^R r'^4 dr' = -\frac{3}{5}\left(\frac{4\pi\rho'}{3}\right)^2 \frac{R^5}{4\pi\varepsilon_0}$$

Inserting the value $\rho' = \frac{Ze}{4/3\ \pi R^3}$, or hence $\frac{4}{3}\pi\rho' = \frac{Ze}{R^3}$, we have

$$W = -\frac{3\ Z^2e^2}{20\pi\varepsilon_0 R} = -\frac{3\ Z^2e^2}{20\pi\varepsilon_0 R_0 A^{1/3}} \tag{8.15}$$

Hence, we have a term for the loss of binding energy due to disruptive Coulomb energy,

$$B_2 = -\frac{3Z^2e^2}{20\pi\varepsilon_0 R_0 A^{1/3}} = -a_c \frac{Z^2}{A^{1/3}} \tag{8.16}$$

where a_c is a coefficient to be determined better from empirical data.

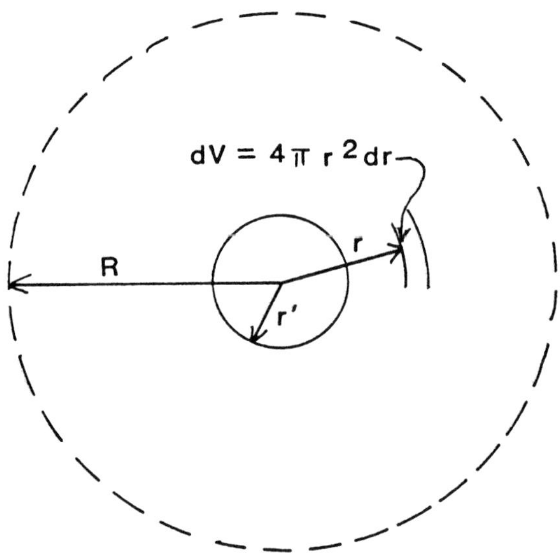

FIGURE 8.2 Geometry for calculating Coulomb energy in a spherical charge distribution.

Another deficit in binding energy depends on the neutron excess (N-Z) and is proportional to $(N-Z)^2/A$, where N is the number of neutrons in the nucleus. This "asymmetry energy" is a purely quantum-mechanical effect. It expresses the physical fact that, in a quantized system of neutrons and protons, any "excess" nucleons are pushed up to energy levels which they occupy alone. They are thus deprived of the fullness of binding. If $|N-Z|$ excess nucleons are regarded as producing a deficit in binding energy because they are "out of reach" of the other nucleons, the fraction of the total nuclear volume so affected is $\frac{|N-Z|}{A}$, and the total deficit is proportional to the product of these, or

$$B_3 = - a_a \frac{(N-Z)^2}{A} = - a_a \frac{(A-2Z)^2}{A} \qquad (8.17)$$

the last equality following from N = A - Z.

There is a non-negligible spin interaction energy which has been neglected in the initial liquid-drop model. This is a simple energy correction term, called the "pairing energy," given by

$$B_4 = \begin{cases} 0 & \text{for odd } A \\ +\delta & \text{for even } Z, \text{ even } N \\ -\delta & \text{for odd } Z, \text{ odd } N \end{cases} \qquad (8.18)$$

where δ is of the order of 0.5 MeV.

Assembling the previous terms, we have the semiempirical mass formula of the liquid-drop model

$$\begin{aligned} M(Z,A)c^2 &= c^2 \, [ZM_H + (A-Z)\, m_n] - B.E. \\ &= c^2 \, [ZM_H + (A-Z)\, m_n] - a_v A + a_s A^{2/3} + a_c \frac{Z^2}{A^{1/3}} \\ &+ a_a \frac{(A-2Z)^2}{A} \pm \delta, \, 0 \end{aligned} \qquad (8.19)$$

Evaluating the coefficients by comparison of data on masses and the energetics of nuclear reactions,

a_v = 14.1 MeV
a_s = 13 MeV
a_c = 0.595 MeV
a_a = 19 MeV

Figure 8.3 shows the separate contributions of each of the energy terms to the B.E./A curve shown previously in Figure 8.1. The initial rise of B.E./A with A is seen to be attributable mainly to the decreasing importance of the surface energy. At larger A the disruptive Coulomb energy becomes dominant, causing a maximum in B.E./A ~ 60 and a subsequent decline in B.E/A at larger A. Throughout the entire range of A above A ~ 40, the semiempirical mass formula matches the observed binding energies to within about 1 percent.

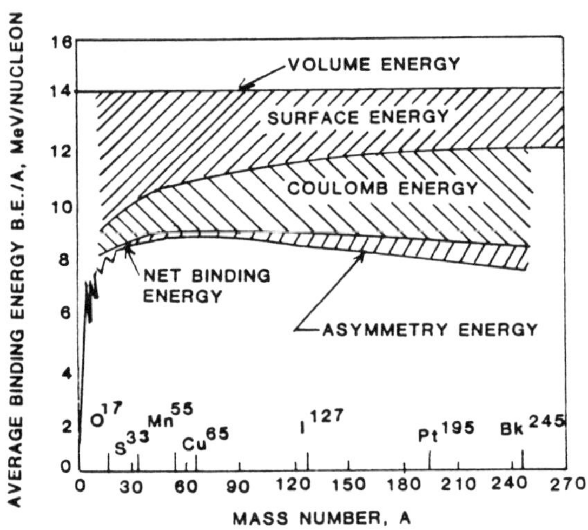

FIGURE 8.3 Summary of the semiempirical liquid-drop-model treatment of the average binding-energy curve. [From Evans, Ref. (8-1)]

Note how the decrease in surface energy and the increase in Coulomb energy conspire to produce the maximum observed in B.E./A at A ~ 60.

With this mass formula, one can now estimate the energy release in the fission process. The division of the nucleus (Z,A) into halves $(\frac{Z}{2}, \frac{A}{2})$ is called symmetric fission. By denoting $M(\frac{Z}{2}, \frac{A}{2})$ as the mass of a fission fragment resulting from symmetric fission, and using Equation (8.19), the expected energy release is

$$
\begin{aligned}
Q &= M(Z, A)\, c^2 - 2M\,(\tfrac{Z}{2}, \tfrac{A}{2})\, c^2 \\
&= a_s\, A^{2/3}\,(1-2^{1/3}) + a_c\, \frac{Z^2}{A^{1/3}}\,(1 - \frac{1}{2^{2/3}}) \\
&= -\,0.260\, a_s\, A^{2/3} + 0.370\, a_c\, \frac{Z^2}{A^{1/3}} \\
&= -\,3.4\, A^{2/3} + 0.22\, \frac{Z^2}{A^{1/3}} \text{ MeV}
\end{aligned}
\qquad (8.20)
$$

For $^{236}_{92}U$,

$$
\begin{aligned}
Q &= -\,3.4\,(236)^{2/3} + 0.22\,\frac{(92)^2}{(236)^{1/3}} \\
&= -\,130 + 301 = 171 \text{ MeV}
\end{aligned}
$$

Note that the positive energy release Q is the result of a large diminution in Coulomb disruptive energy (301 MeV), which overrides a smaller change in the opposite direction (-130 MeV) due to the increased ratio of surface to volume. Thus the energetics of nuclear fission is seen to depend mainly on the interplay between Coulomb energy and surface energy.

We define the neutron separation energy S_n as the work required to remove the last neutron from a nucleus which contains N neutrons and Z protons. Then

$$
\begin{aligned}
S_n(Z,N) &= [M(Z,N-1) + M_n - M(Z,N)]\, c^2 \\
&= B.E.(Z,N) - B.E.(Z,N-1)
\end{aligned}
\qquad (8.21)
$$

The amount of energy needed to overcome the nuclear potential barrier is relatively small (~ 5 MeV in the heaviest nuclides). This excitation energy can be supplied in many ways to induce fission. For example, immediately after the capture of a thermal neutron, the compound nucleus has an excitation energy of S_n, the neutron separation energy from the ground level of the compound nucleus. This excitation energy can be released, for example, by γ-ray emission, or the excited nucleus will fission. A study of S_n by use of Equation (8.19) shows that S_n is appreciably greater for even-N than for odd-N nuclides as a consequence of the pairing energy δ. Consider the excitation energy produced when a thermal neutron is captured by an odd-N target (say $^{235}_{92}U$) compared with an even-N target (say $^{238}_{92}U$) of approximately the same mass number. If we insert (8.19) into (8.21), the excitation energy in

$$^{235}_{92}U + n \rightarrow {}^{236}U$$

is substantially $2\delta \sim 1$ MeV greater than the excitation energy in the compound nucleus formed by

$$^{238}_{92}U + n \rightarrow {}^{239}U$$

The fact that ^{235}U undergoes fission with thermal neutrons (~ 0.025 eV), but ^{238}U requires bombardment by fast neutrons (~ 1 MeV) in order to undergo fission is attributed to this expected difference of 2δ in the excitation energies.

A summary of what causes fission is as follows. The competing effects of surface energy and Coulomb energy lead to a potential barrier with a barrier height ~ 5 MeV for heavy nuclides. A neutron added to the nucleus disrupts the surface energy by S_n to the point where the nucleus is excited above the potential barrier. Fission is then one method of de-excitation.

8.1.2 Charateristics of the fission process. Although in the previous section, we discussed symmetric fission, in actuality heavy nuclides fission asymmetrically with thermal and fast (~1 MeV) neutrons, producing a spectrum of fission products, with two distinct peaks: a "light" group with most probable mass number of about 95 and a "heavy" group with most probable mass number of about 138. Typical values of the nuclear charge numbers of the most probable light and heavy fragments are Z = 38 (Strontium) and Z = 54 (Xenon). This asymmetry is not well understood. A typical yield curve for fission of ^{235}U is shown in Figure 8.4. Fission yield is expressed such that the sum of all yields is 200 percent as expected theoretically for binary fission.* Note an interesting feature in the curve by comparing the thermal neutron (~0.025 MeV) and 14-MeV neutron yields. Presumably at extremely high energy, symmetric fission will be the most probable mode. However, for reactor purposes there is not much difference on this scale between a thermal neutron fission and fast (1-MeV) neutron fission, so the curve may be taken as universal for ^{235}U.

Many of the fission products have fairly low boiling points. If we assume at 1 atm pressure that these products become gaseous, we can construct a fission gas yield curve as a function of temperature. The data are shown in Table 8.1 and shown graphically in Figure 8.5 [Reference (8-2)]. These are minimum estimates since boiling point data are lacking on many of the fission products. On the other hand, considering all parameters (time, flux, burnup, decay) one can probably do no better at this time without exhaustive calculations. Fairly universal agreement appears

---

* Only about 1 in every 100,000 events gives three fission fragments. One out of about 400 results in an energetic alpha particle plus two "regular" fission fragments.

for the 25% generation figure above 700°C, but above 1100°C there is considerable controversy.

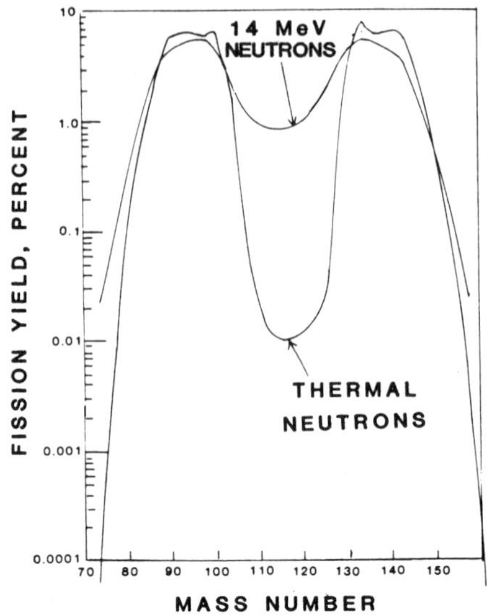

FIGURE 8.4 Yield-mass curve for fission of ^{235}U.

[Permission granted by the University of California, Lawrence Berkeley Laboratory to reproduce from Earl K. Hyde, "A Review of Nuclear Fission, Part One-Fission Phenomena at Low Energy," UCRL-9036, January, 1960]

TABLE 8.1

Fission Gas Yield for Fission of ^{235}U

B.P. °C	Gas	Yield	Cumulative Yield	Cumulative Percent of Fission Fragments
-152.9	Kr	3.56	3.56	1.78
-107.1	Xe	21.74	25.30	12.65
58.8	Br	0.15	25.45	12.72
184.4	I	1.03	26.48	13.24
670.0	Cs	19.15	45.63	22.81
700	Rb	3.79	49.42	24.71
1140	Ba	5.74	55.16	27.58
1150	Sr	14.13	69.29	34.64
1390	Te	2.0	71.29	35.64
1490	Y	5.4	76.69	38.34

[From Hyde, Ref. (8-2)]

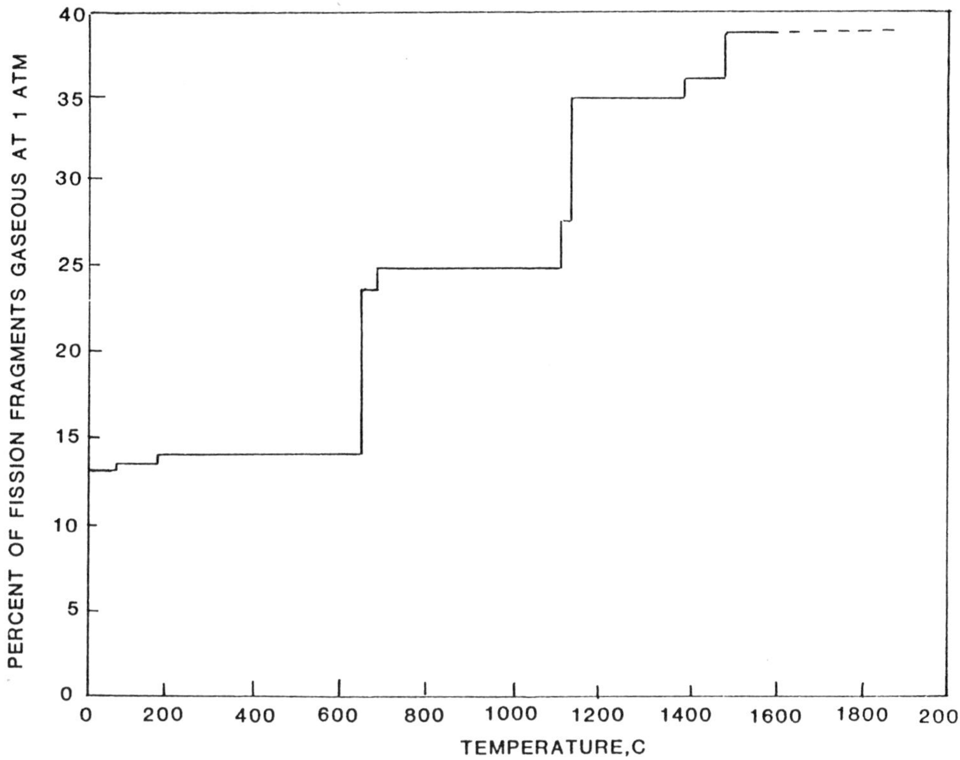

FIGURE 8.5 Fission gas generation at 1 atm versus temperature. Minimum estimate, for ~1 year operation.

As with the distribution in mass number of the fission fragments, there is also a bimodal curve for the energy distribution, as shown in Figure 8.6. The most probable energies for the light and heavy fragments are 98 and 67 MeV. The experimental N(E) distribution can be approximated by the expression [Reference (8-3)]

$$N(E) = 0.072[0.685e^{-\alpha(E-67)^2} + e^{-\delta(E-98)^2}] \quad 0 \le E \le E_0 \tag{8.22}$$

where $\alpha = 0.00693$, $\delta = 0.0177$, $E_0 = 115$ MeV, and this function has been normalized to 2 fragments per fission.

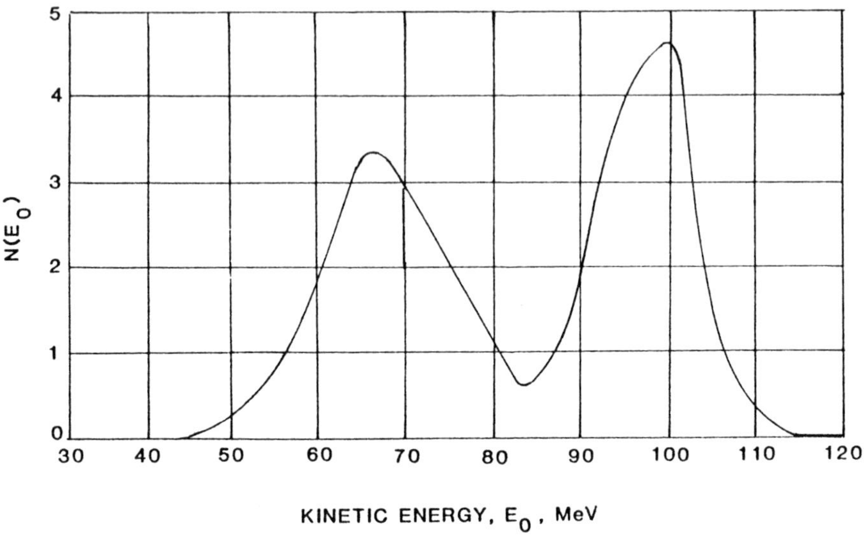

FIGURE 8.6 Energy distribution of fission fragments from uranium-235.

The initial distribution of the energy per fission is approximately as given in Table 8.2 [Reference (8-4)]. Note that the instantaneous energy release per fission is (fragment kinetic energy) + (neutron kinetic energy) = 168 + 5 = 173 MeV, and compare this with the value of 171 MeV calculated previously. So far as sensible heat released in the fission process, within a very short time Items 1, 2, 4, and 5 (of Table 8.2) = 183 MeV is released. Neutrinos are not captured in the vicinity of the fuel (and many not even by the earth!), so this energy (10 MeV) never appears as sensible heat. The remaining energy, of betas and gammas (11 MeV) from decay of the fission products, is gradually released, so that after long time operation of a reactor, the total heat release per fission is 183 + 11 = 194 MeV.

TABLE 8.2

Energy per Fission for ^{235}U

Item	Energy, MeV
1. Kinetic energy of the fragments	168
2. Internal excitation of the fragment nuclei	5
3. Energy released in beta decays	
a. beta particles	6
b. gamma rays	5
c. neutrinos	10
4. Kinetic energy of neutrons	5
5. Energy released when neutrons are captured	5
Total	204

Now consider the range of the fission fragments in a material. One comes across various definitions for the range of a particle in a material, but the most physically significant parameter is the "mean range," and this is used in all range-energy relationships. The "stopping power" of a material is defined as $(\frac{dE}{dx})$, where dx is an element of path. The range R is then given from the range energy relationship

$$E = \int_0^R (\frac{dE}{dx})\, dx \qquad (8.23)$$

However, rather than introduce undue complexity, let us assume that $(\frac{dE}{dx})$ is linear over the range. This approximation for fission fragments is illustrated in Figure 8.7 [Reference (8-1)]. The individual range of many of the fission fragments in uranium metal has been determined experimentally [Reference (8-2), Table 11.26]. These data are shown graphically in Figure 8.8 and indicate approximately a linear behavior of range with fission fragment mass number. Also shown in this graph is the product of the individual fission yields and ranges which emphasizes the two-group behavior of the fragments. The ratio of the average range of the light fragment to the average range of the heavy fragments is 1.29. The mean range of fission fragments in UO_2 is 9.4 microns [Reference (8-5)]. Applying the 1.29 ratio the light fragment range is 10.6 microns and the heavy fragment range is 8.2 microns. The mass range is $R_n = \rho R = 9.5\ mg/cm^2$ for 92 percent of the theoretical density, $10.96\ gm/cm^3$. The Bragg-Kleeman empirical rule states that the mass range is proportional to the square root of the atomic weight, so that knowing the range for UO_2, we can find it for any fuel. (Note, the effective atomic weight for a compound for use in the Bragg-Kleeman rule is $\sqrt{A} = \sum_i \gamma_i A_i / \sum_i \gamma_i \sqrt{A_i}$ where γ_i is the number of atoms of the ith kind in the compound and A_i is the atomic weight of the ith species.) Knowing the mean range of the two groups in UO_2 and the energy per fission fragment we can arrive at a mean value of the stopping power for model calculations, i.e., $(\overline{\frac{dE}{dx}}) = \frac{E_0}{R}$. For the individual groups in UO_2,

$$(\overline{\frac{dE}{dx}})_{light} = \frac{98}{10.6 x 10^{-4}} = 9.25 \times 10^4 \text{ MeV/cm}$$

$$(\overline{\frac{dE}{dx}})_{heavy} = \frac{67}{8.2 x 10^{-4}} = 8.18 \times 10^4 \text{ MeV/cm}$$

$$(\overline{\frac{dE}{dx}})_{average} = \frac{168}{(2)(9.4 x 10^{-4})} = 8.95 \times 10^4 \text{ MeV/cm}$$

The range of fission fragments in nuclear fuels of typically 10 microns is generally sufficiently small to consider the fission energy release to be at the point of fission, for engineering calculations.

8.2 Effects in Reactor Fuel Elements

Now that the fission process has been described, consider the effects of this process on the interior of reactor fuel elements.

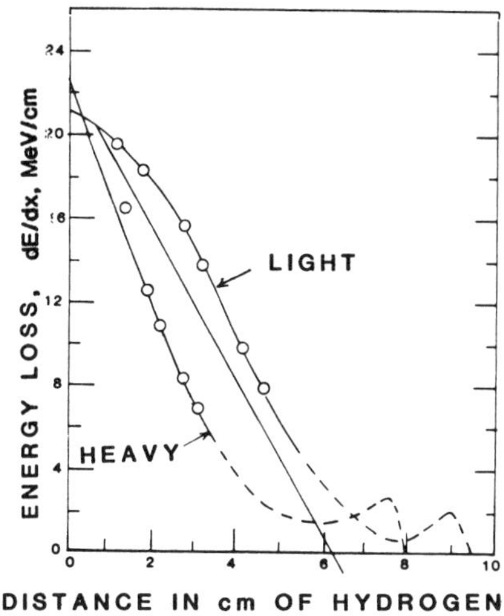

FIGURE 8.7 Energy loss per unit path length for average heavy and light fission fragments from ^{235}U (n,f).

[Reproduced with permission from the American Physical Society; N.O. Lassen, "On the Energy Loss by Fission Fragment Along Their Path," Phys. Rev., 75, p. 1762, 1949]

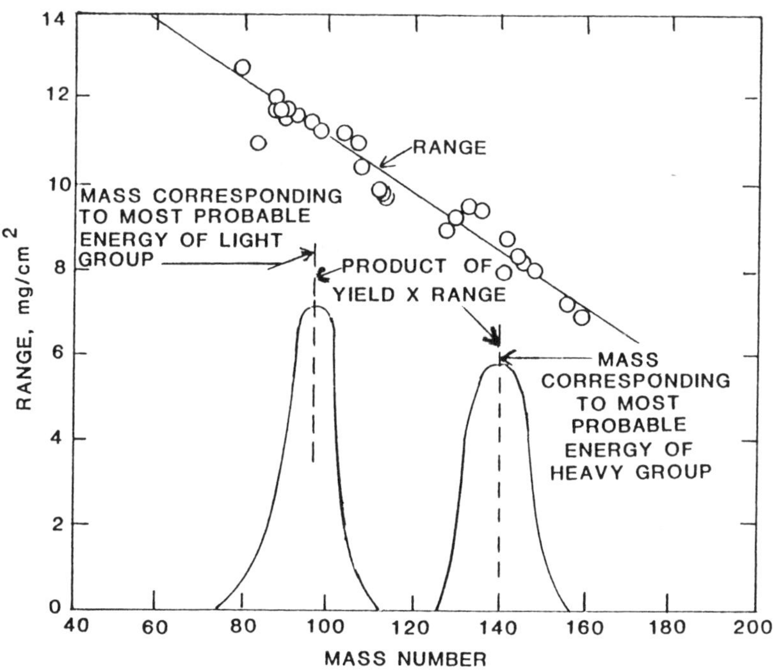

FIGURE 8.8 Range of fission fragments in uranium metal. [From Hyde, Ref. (8-2)]

8.2.1 Thermal spike. When a charged particle interacts with a material, its kinetic energy is converted ultimately into thermal energy in a very short distance. This local deposition of energy in a small volume, or microvolume, with the resulting elevation of local temperature is referred to as a "thermal spike."

In a mathematical model of the thermal spike, it is imagined that the stopping of the charged particle produces essentially an instantaneous line source of thermal energy at infinite temperature [Reference (8-6)]. That the source is essentially instantaneous can be illustrated by example.

Illustration 8.1 Estimate the duration of time to form a thermal spike. Consider an "average" fission fragment with mass of 117 amu, initial kinetic energy of 84 MeV, and range R in UO_2 of 9.4×10^{-6}m. Take the average velocity while being slowed down as $\bar{v} = 1/2 \sqrt{\frac{2E}{m}} = 0.59 \times 10^7$ m/s. Thus the time to form a thermal spike is approximately $t = R/\bar{v} = 1.6 \times 10^{-12}$ sec.

It is then assumed that the heat diffuses radially from the line source, elevating the temperature of the surrounding material. The gross, and perhaps unrealistic, assumption is made that macroscopic concepts and physical properties are applicable on a very short time scale and in a volume containing relatively few atoms. Nevertheless, such assumptions are necessary to render the problem tractable for a thermal calculation.

Fourier's heat conduction equation in general form is

$$\rho c \frac{\partial T}{\partial t} = \nabla \cdot (k \nabla T) \tag{8.24}$$

where

T = temperature

t = time

ρ = density of the material

c = specific heat

k = thermal conductivity

Using cylindrical coordinates along the straight-line path of the charged particle (X-axis), assuming all heat to be conducted radially and with constant physical properties, Equation (8.24) reduces to

$$\frac{\partial T}{\partial t} = \frac{\alpha}{r}\frac{\partial}{\partial r}\left(r\frac{\partial T}{\partial r}\right) \tag{8.25}$$

where α is the thermal diffusivity ($\alpha = \frac{k}{\rho c}$). Assuming initial conditions at time t = 0 of

$$\begin{aligned} T &= 0 \text{ for } r \neq 0 \\ T &= \infty \text{ for } r = 0 \end{aligned} \tag{8.26}$$

Equation (8.25) has as its solution (neglecting latent heat*)

$$T\,(r,t) = \frac{C_1}{t}\, e^{-\frac{r^2}{4\alpha t}} \tag{8.27}$$

where C_1 is a constant with respect to r and t evaluated from the assumption that all heat is conducted radially, i.e., from the general law of heat diffusion that the heat flux q' = $-k\nabla T$, we have for this case

$$\left[\frac{dE}{dX}\right] = -2\pi k \int_0^\infty r\,\frac{\partial T}{\partial r}\,dt \tag{8.28}$$

---

* For a material such as UO_2, this assumption is not a severe limitation to the theoretical development. For example, the specific enthalpy of UO_2 at the melting point is more than four times the latent heat of fusion [Reference (8-7)].

where $\left[\frac{dE}{dX}\right]$ is the stopping power (energy loss per unit path in the material). From Equations (8.27) and (8.28)

$$C_1 = \frac{\left[\frac{dE}{dX}\right]}{4\pi k}$$

Thus, the final temperature solution is

$$T(r,t) = \frac{\left[\frac{dE}{dX}\right]}{4\pi kt} e^{-\frac{r^2}{4\alpha t}} \tag{8.29}$$

In this expression, $\left[\frac{dE}{dX}\right]$ can be, in general, a weak function of X, so long as the axial heat conduction is negligible with respect to the radial conduction.

To restrict the thermal spike to be finite in volume and finite in duration, consider the spike to have influence only at temperatures above the melting point of the material. Now at any given radius ($r \neq 0$), the temperature first increases to some maximum temperature and then decreases with time. The maximum temperature is found from

$$\frac{\partial T}{\partial t} = \frac{T}{t}\left(\frac{r^2}{4\alpha t} - 1\right) = 0 \tag{8.30}$$

to be

$$T_{max} = \frac{\left[\frac{dE}{dX}\right]\alpha}{\pi k e r^2} \tag{8.31}$$

where e is the base of the Naperian logarithm. At radii at which the temperature reaches the melting temperature T_m (above ambient), we thus have the "boundary" of the thermal spike from Equation (8.31) as

$$r_0 = \left\{\frac{\left[\frac{dE}{dX}\right]}{\pi e \rho c T_m}\right\}^{1/2} \tag{8.32}$$

From Equation (8.30), the duration t_0 of the spike can thus be approximated as

$$t_0 = \frac{r_0^2}{4\alpha} = \frac{\left[\frac{dE}{dX}\right]}{4\pi e k T_m} \tag{8.33}$$

The microvolume V contained in the spike can readily be obtained from Equation (8.32) as

$$V = \int_0^{R_p} \pi r_0^2 dX = \frac{E}{e c \rho T_m} \tag{8.34}$$

where R_p is the range of the charged particle and E is the particle initial energy. Note that the microvolume size is independent of the thermal conductivity. If one allows for a temperature dependency of the density and specific heat, the microvolume size is slightly altered to [Reference (8-6)]

$$V = \frac{E}{2c_m \rho_m T_m} \tag{8.35}$$

where c_m and ρ_m are the specific heat and density at the melting temperature.

<u>8.2.2 Fuel swelling</u>. One effect of the fission process in nuclear fuel is that of swelling. This phenomenon also presents an example of the application of the thermal spike model in radiation effects. Fuel elements show an increase in dimensions during irradiation. This is caused by accumulation of fission product atoms in the structural lattice of the fuel. Some of these atoms (when grouped) are gases. Others are solid at the given temperature of fuel operation. The solid fission products occupy a larger volume than the original fuel and cause solid volume swelling. However, the primary source of swelling is from the noble gases (e.g., Kr, Xe) [See Table 8.1] which nucleate and

grow into bubbles. The gas atoms formed by fission diffuse within the lattice of the fuel until enough atoms come together to form a stable nucleus. Dislocations can possibly serve as nuclei. Gas solubility is assumed so small that once a gas atom enters a bubble, it will not leave. The bubble grows (swells) as gas atoms diffuse into it. Even small bubbles may diffuse through the system and coalesce into large ones (generally at grain boundaries). There are many swelling models available. Two of them will be considered here.

The first model considers the swelling to occur as a result of plastic deformation of the fuel material. It is assumed at the start that there are N'sites per unit volume for nucleation of gas bubbles. For descriptive purposes, assume each of these sites is a small spherical cavity with radius r_0, although this assumption will later drop out of the problem. Now, the total number of gas atoms R generated per unit volume at time t after the start of the irradiation is derived from the simple reaction rate equation to be

$$R = \Sigma_f \mu \Phi t \tag{8.36}$$

where

Σ_f = macroscopic fission cross section
μ = number of gas atoms generated per fission
Φ = average neutron flux

However, for long time operation of the fuel, the number of fuel atoms (hence, the macroscopic fission cross section) will decrease. If N is the number of fissionable nuclei per unit volume, then the number "burned up" in time dt is

$$dN = -\sigma_a \Phi N dt \tag{8.37}$$

where σ_a is the microscopic absorption cross section of the fuel, so that

$$N = N_0 e^{-\sigma_a \Phi t} \tag{8.38}$$

where N_0 is the initial concentration at time $t = 0$. The gas generation rate is thus

$$\frac{dR}{dt} = \Sigma_f \Phi \mu = \sigma_f N \Phi \mu = \Phi \mu \sigma_f N_0 e^{-\sigma_a \Phi t} \tag{8.39}$$

Integrating Equation (8.39)

$$R = \mu \Sigma_{f0}(1 - e^{-\sigma_a \Phi t})/\sigma_a \tag{8.40}$$

where $\Sigma_{f0} = N_0 \sigma_f$ and N_0 is the initial number of fissionable fuel atoms. This equation replaces the linear approximation of Equation (8.36). The ideal gas law is assumed to hold in each cavity

$$pV = nKT \tag{8.41}$$

where

p = pressure in each cavity
V = volume of cavity
n = total number of gas atoms in each cavity
K = gas constant
T = absolute temperature

Combining Equations (8.40) and (8.41), we have

$$p = \frac{\mu \Sigma_{f0}(1-e^{-\sigma_a \Phi t})\ KT}{\sigma_a N' \ (\frac{4}{3}\ \pi r^3)} \tag{8.42}$$

where

r = radius of cavity at time t.

The problem now is to determine the pressure in each cavity (or hence the radius r of each cavity). It is assumed that as the pressure builds up in each cavity, the material adjacent to the cavity is deformed plastically. The cavity expands, resulting in fuel element swelling.

The following assumptions are made for the deformation analysis:

1. The cladding does not restrain the fuel material from swelling.

2. The fuel material can be treated as an isotropic solid with $\gamma/E \ll 1$. (γ = yield strength, E = Young's modulus)

3. During plastic deformation, there is no work-hardening effect.

4. The material does not creep for the stress states of interest.

5. The stress fields of adjacent cavities do not interact.

6. The rate of deformation is such that the problem can be treated as a quasi-stationary case.

7. Surface tension is negligible.

8. Von Mises' yield criterion applies.

Then, with the final radius of the gas bubbles assumed much larger than the initial radius, the solution for the pressure inside each cavity is [Reference (8-8)]

$$p = \frac{2}{3}\gamma\{\ln\left[\frac{E}{3\gamma(1-\nu)}\right] + 1\} \qquad (8.43)$$

where ν is Poisson's ratio. Combining Equations (8.42) and (8.43) results in the volume growth for a single cavity of

$$\frac{4}{3}\pi r^3 = \frac{3\mu}{2}\,\frac{\Sigma_{f0}\,(1-e^{-\sigma_a\Phi t})\,KT}{\gamma\{1+\ln\,[\frac{E}{3\gamma(1-\nu)}]\}\,\sigma_a N'} \qquad (8.44)$$

And since there are N' cavities per unit volume, the change in volume per unit volume can be written as

$$\frac{\Delta V}{V} = N'\frac{4}{3}\pi r^3 = \frac{3\mu KT\,\Sigma_{f0}\,(1-e^{-\sigma_a\Phi t})}{2\gamma\sigma_a\{1+\ln\,[\frac{E}{3\gamma\,(1-\nu)}]\}} \qquad (8.45)$$

But, the fractional atom burnup is [from Equation (8.38)]

$$\frac{N_0-N}{N_0} = 1-e^{-\sigma_a\Phi t} = \frac{\Delta N}{N_0} \qquad (8.46)$$

so that the equation for swelling can be written directly in terms of burnup as

$$\left(\frac{\Delta V}{V}\right) = \frac{3\mu KT\quad\Sigma_{f0}}{2\gamma\sigma_a\;\{1+\ln[\frac{E}{3\gamma(1-\nu)}]\}}\left(\frac{\Delta N}{N_0}\right) \qquad (8.47)$$

A correction has yet to be made for the contribution to swelling due to the solid fission fragments. For UO_2, this can be shown to be 0.75ε percent per a/o burnup [Reference (8-9)] where ε is the fractional enrichment.* Thus, the final corrected equation becomes

$$\frac{\Delta V}{V}\,(\%) = [\frac{3KT\mu\Sigma_{f0}}{2\gamma\sigma_a\;\{1+\ln\,[\frac{E}{3\gamma(1-\nu)}]\}}] + 0.75\varepsilon]\;\frac{\Delta N}{N_0}(\%) \qquad (8.48)$$

---

* For UC fuel, the solid swelling is 1.01ε percent per a/o burnup.

One of the problems encountered when comparing theoretical results with experimental results is an accurate knowledge of the fuel temperature. This has been carefully analyzed for four UC specimens [Reference (8-9)]. Table 8.3 presents a comparison of Equation (8.48) with experimental results. For this comparison, an in-pile yield strength of 3,000 psi was used for cast UC. As seen from this table, the simplified formula of Equation (8.48) predicts the swelling magnitude fairly well. However, the out-of-pile yield strength of UC is considerably greater than the value assumed to match the experimental results. It must therefore be concluded that either the yield strength is reduced in a fissioning environment, or the plastic deformation model underestimates the swelling. That the yield strength could be reduced by radiation is counter to most observations (for example, see Figure 7.7).

TABLE 8.3

Comparison of Theory and Experiment for Swelling of Irradiated UC

Specimen	$(\frac{\Delta V}{V})$ % Measured	$(\frac{\Delta V}{V})$ % Equation (8.48)
113	5.0	4.0
138	3.2	3.2
140	4.2	3.7
141	4.8	3.2

A strong objection to the plastic deformation model is that it predicts a linear temperature-dependency of the swelling, whereas observations indicate that at relatively high temperatures "breakaway swelling" occurs in which the swelling exceeds a linear curve with temperature. Unless the in-pile yield strength exhibits a strong temperature dependency, the plastic deformation model must be abandoned in favor of other models.

One model now favored is to consider swelling to be due to creep of the fuel material. A typical creep curve is sketched in Figure 8.9. We normally consider creep in the secondary creep mode. Here the slope, i.e., creep rate is approximately constant. For UO_2

$$\dot{\varepsilon} = K\sigma^4 e^{-40,000/T} \tag{8.49}$$

where

$\dot{\varepsilon}$ = strain rate
σ = applied stress
T = temperature, °K

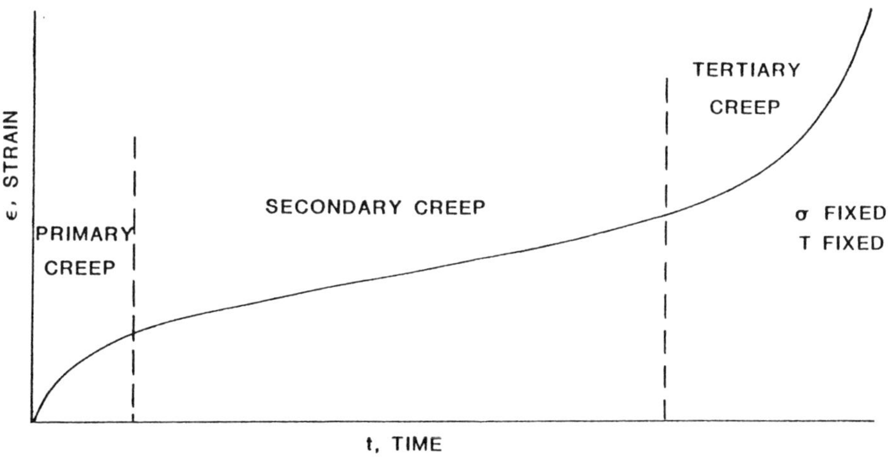

FIGURE 8.9 Sketch of typical creep behavior.

A more general form of this creep law is written as

$$\dot{\varepsilon} = \frac{1}{B(T)} \sigma^n \tag{8.50}$$

Freshly annealed material will start out in primary creep, but soon work hardens and behaves in the secondary creep mode. We now consider the case where we have a gas bubble with pressure (stress) p inside. The surrounding material creeps under the stress of the bubble pressure and, hence, the fuel element swells. Analytically, the problem is much more complex than the simple plastic deformation model considered previously.

The pore pressure p is related to the gas generation rate, the instantaneous pore radius, the temperature, and the degree of gas released. The ideal gas law has been used as the equation of state of the gas; hence, p can be replaced by the above-mentioned parameters in a relatively straight-forward manner. The model [Reference (8-10)] may be employed to predict the behavior of cermet-dispersion fuels or to describe the behavior of homogeneous alloy fuels. In the analysis of swelling of cermets, the fuel particles are assumed to have no strength, to only occupy space (i.e., the fuel strength is considered negligible in comparison with the matrix material), and to act as sources of gas that exert pressure on the matrix. Therefore, the restraining members of the system are the matrix and the cladding. In the analyses for homogeneous fuels, however, pressure is assumed to build up in bubbles within the fuel; then, the restraining members are the fuel and the cladding. A particular case considered corresponds to the swelling of a 0.2-in.-o.d. fuel pellet surrounded by a 20-mil cladding at a constant temperature of 1700°C. The fuel is 99% dense and has a yield stress of 5,000 psi and a creep strength* of 70 psi. This creep strength was

---

* Creep strength is defined as the stress to produce a steady-state strain rate of 10^{-6}/hr.

chosen because it best fits the experimental swelling data for UC. As shown in Figure 8.10, [Reference (8-10)], the creep strength, chosen so that the calculations will approximate the swelling behavior of UC at 1700°C, is about one third that of the out-of-pile creep strength of UC. The required reduction in the creep strength of UC is assumed to be the result of in-pile fission effects on creep. Other parameters for the calculations are a cladding yield stress and creep strength of 5,000 and 1,000 psi, respectively. The variation of strain rate with stress is assumed to be to the 5.3 power for the cladding and to the 4 power for the fuel. Finally, a fission rate of 1.3 x 10^{13} fissions/(cm^3)(sec) and no gas release are assumed.

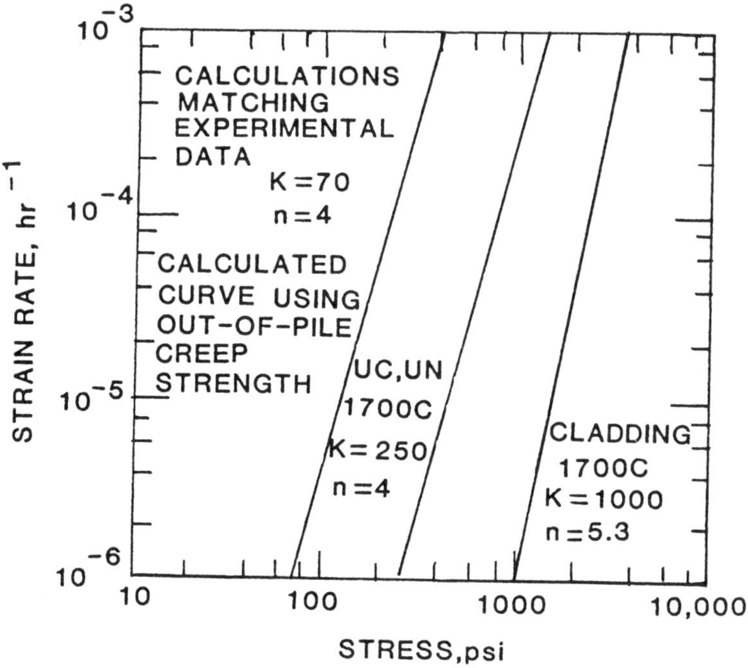

FIGURE 8.10 Stress vs. strain rate for fuel and cladding (K = creep strength).

[Reproduced with permission from J.A. Walowit, W.F. Adler, and W. Chubb, "Mechanical Analysis of Fission-Gas Swelling of Clad Cylindrical Fuel Elements," Reactor and Fuel - Processing Technology, 12, No. 2, 1969]

As mentioned previously, in order to match the experimental data, it must be assumed that the in-pile creep strength is considerably less than that determined from out-of-pile measurements. The model to explain this behavior is as follows. As discussed earlier in this chapter, each fission produces two fission fragments. These fragments possess initially about 168 MeV total. In a very short distance (~10μ = 10^{-5}m = 10^{-3}cm), the fragments are stopped in the material, and their kinetic energy is converted to heat. These localized thermal spikes are analyzed in the previous section and locally anneal the material, so that the local material reverts back to the primary mode of creep. In a predictable time, t_m, essentially all the fuel is annealed once and the process repeats. The situation is sketched in Figure 8.11.

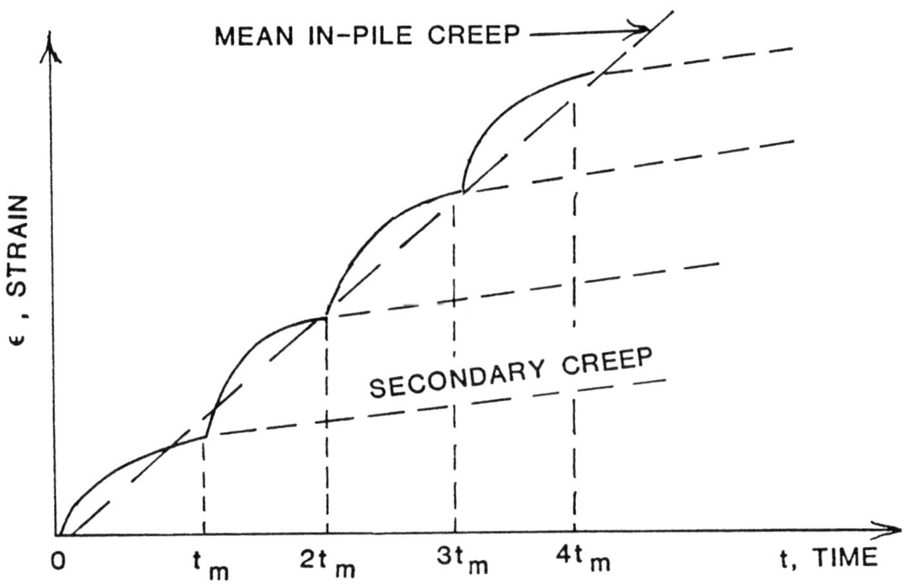

FIGURE 8.11 The effect of annealing on creep rate.

From Equation (8.35), the melted microvolume for a thermal spike is given as

$$\Delta V = \frac{E_F}{2ec_v \rho T_m}$$

where E_f is the kinetic energy per fission. The production of fission fragments (neglecting burnup) is $N = 2\Sigma_f \Phi Vt$, where Φ is the neutron flux, Σ_f is the macroscopic fission cross section, and V is the fuel volume, so that the ratio of melted volume to fuel volume is

$$\eta = \frac{N\Delta V}{V} = \frac{E_f \ \Sigma_f \Phi \ t}{ec_v \ \rho \ T_m} \qquad (8.51)$$

[The process is really a statistical (Poisson) mapping, but Equation (8.51) is a rough approximation.] At this point, it is instructive to estimate the radiation exposure required for $\eta = 1$, i.e., when essentially all the fuel volume has undergone melting once. Assuming UO_2 fuel, $\eta = 1$ when $\Sigma_f \Phi t \simeq 10^{15}$ fissions/cm^3(100 seconds irradiation time at a fission rate of 10^{13} fissions/(cm^3)(sec)). Thus in an in-pile irradiation of significant exposure, the melted volume is "turned over" many times.

Experimental results of this creep model are given in Figure 8.12 for the case of the estimated in-pile creep strength of UC at $\Sigma_f \Phi = 7 \times 10^{13}$ fissions/(cm^3)(sec) based on swelling data. In this case, the "turnover" time t_m= 7 sec. Data for primary creep are lacking for very short times after the stress is applied. Based upon the experimental primary creep rates at 20 and 60 minutes, a value at 7 seconds was determined. The agreement of the 7-second calculation with the in-pile creep curve is seen to be good. Note that use of secondary creep grossly underestimates the strain rate.

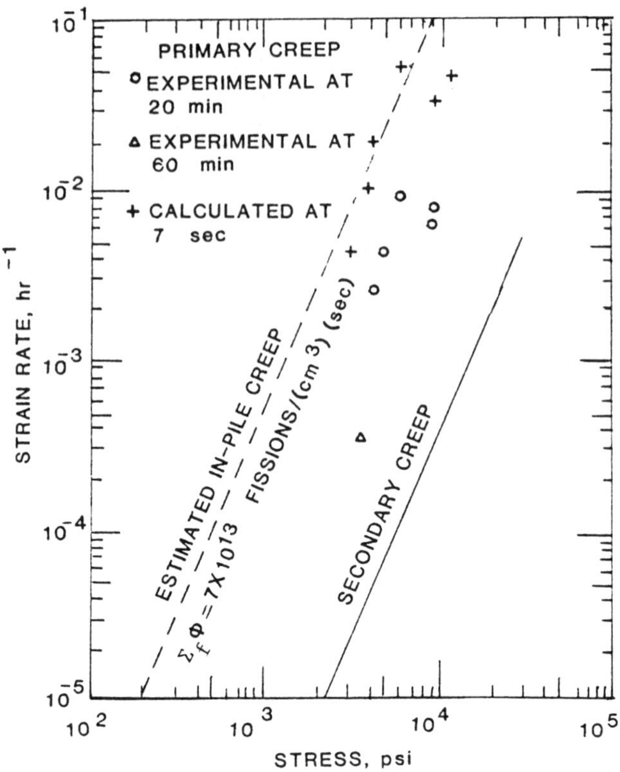

FIGURE 8.12 Comparison of in-pile creep and out-of-pile primary creep of UC at 1300°C.

REFERENCES

(8-1) Evans, R.D., The Atomic Nucleus, McGraw-Hill Book Company, Inc., 1955, Chapter 11.

(8-2) Hyde, Earl K., "A Review of Nuclear Fission, Part One - Fission Phenomena at Low Energy," UCRL-9036, January, 1960.

(8-3) Redmond, R.F., Klingensmith, R.W., and Anno, J.N., "Energy Spectrum of Fission Fragments Emitted from Thin Layers of Uranium Dioxide," J. Appl. Phys., 33, 3383-3385, November, 1962.

(8-4) Wheeler, J.A., Chapter on Fission in *Handbook of Physics*, Edited by E.U. Condon and Hugh Odishaw, McGraw-Hill Book Company, New York, 1958.

(8-5) Holden, A.N., *Physical Metallurgy of Uranium*, Addison-Wesley Publishing Company, Inc., 1958, p. 206.

(8-6) Kalimullah and Anno, J.N., "Thermal Spike Model with Temperature-Dependent Specific Heat," *Transactions of the ASME*, *Journal of Heat Transfer*, *95*, pp. 551-553, 1973.

(8-7) Kerrick, J.F., and Clifton, D.G., "Smoothed Values of the Enthalpy and Heat Capacity of UO ," *Nucl. Technology*, *16*, pp. 531-535, 1972.

(8-8) Hill, R., *The Mathematical Theory of Plasticity*, Oxford Press, pp. 98-101, 1950.

(8-9) Leach, E.R., Ray, J.W., and Dingee, D.A., "Analytical Prediction of Irradiation Swelling of Uranium Carbide Under Mechanical Restraint," Report BMI-1743, Battelle Memorial Institute, Columbus, Ohio, September, 1965.

(8-10) Walowit, J.A., Adler, W.F., and Chubb, W., "Mechanical Analysis of Fission-Gas Swelling of Clad Cylindrical Fuel Elements," *Reactor and Fuel-Processing Technology*, *12*, No. 2, 1969.

PROBLEMS FOR CHAPTER 8

1. For the thermal spike model

   a. Show that $T(r,t) = \frac{C_1}{t} e^{-r^2/4\alpha t}$ is a solution of Fourier's equation

   $$\frac{\partial T}{\partial t} = \frac{\alpha}{r} \frac{\partial}{\partial r} \left(r\frac{\partial T}{\partial r}\right)$$

   b. Show that

   $$C_1 = \left[\frac{dE}{dX}\right] / 4\pi k$$

   c. Show that

   $$\frac{\partial T}{\partial t} = \frac{T}{t} \left(\frac{r^2}{4\alpha t} - 1\right)$$

2. Estimate the average "turnover" time for Zimmer fuel.*

3. Estimate the minimum swelling (solid swelling) in the Zimmer UO_2 fuel, 3 months later after startup.

---

* Zimmer core data:

Average thermal flux = 3.5×10^{13} n/(cm^2)(sec)
Average volumetric fuel temperature = 1100°F
Melting temperature = $5080 - 63.5 \times 10^{-4}$E °F
(where E = Exposure MWd/Ton U, assume 10,000 in this problem)
UO_2 density = 95% theoretical
Average enrichment = 1.9%
Specific heat = 0.078 cal/gm°C

9 SURFACE EFFECTS OF NUCLEAR RADIATIONS

An often neglected area of radiation effects is the interaction of nuclear radiations with surfaces. Most of the discussion of this chapter deals with metallic surfaces, but the effects described are more general. Three principal effects are discussed in this chapter: fast neutron erosion (sputtering), secondary electron emission, and radiation-induced outgassing. For a more complete discussion of the interaction of ions with surfaces, the reader is referred to Reference (9-1).

9.1 Fast Neutron Erosion

When fast neutrons strike a surface, they eject surface atoms. This process, one of erosion, is termed fast neutron "sputtering." The yield, or sputtering ratio, is defined as

$$\Delta_T = \frac{\text{number of atoms ejected from a surface}}{\text{number of fast neutrons crossing the surface}}$$

This number has engineering significance not only to erosion of fuel cladding in high flux fission reactors, but also to fusion reactor vessels in which neutrons are produced by the neutron branch of the D-D reaction and by the D-T reaction.

To predict the fast neutron sputtering ratio, consider the following. The reaction rate R of fast neutrons with a material is given by (see Chapter 3)

$$R = \Sigma_s \Phi = N \sigma_s \Phi \tag{9.1}$$

where

Σ_s = macroscopic scattering cross section
N = atomic density of material
Φ = fast neutron flux
σ_s = microscopic scattering cross section

In general, the energy exchange will be such that the scattering nucleus (atom) is ejected from its lattice site. Assuming its path to be a straight line and its range in the material to be B, then the number of scattered atoms leaving the surface is

$$S = RB/4 \tag{9.2}$$

(The 1/4 factor comes from the fact that in an isotropic production medium, 1/2 of the atoms will be directed inwardly and 1/2 of those directed outwardly will arrive at the surface - see Appendix B.) Now the number of fast neutrons crossing the surface from both sides is $\Phi/2$. Thus the sputtering ratio calculated in this manner is

$$\Delta = \frac{2S}{\Phi} = \frac{RB}{2\Phi} = \frac{\Sigma_s B}{2} \tag{9.3}$$

The range B of the scattered atoms in the material can be determined from the classical stopping power formula. The average initial energy E_p of the scattered atoms, or primary knock-ons, is given for an elastic scattering event as (see Chapter 3)

$$\overline{E}_p - \frac{2m_n m_T}{(m_n + m_T)^2} E_n \tag{9.4}$$

Where

m_n = mass of neutron
m_T = mass of target atom
E_n = fast neutron energy

The predictions of Equation (9.3) must be enlarged due to the fact that the PKAs will in turn dislodge other atoms from their sites. A rough estimate of the number of atoms ν displaced is, from Chapter 3,

$$\nu = \frac{\overline{E}_p}{2E_d} \tag{9.5}$$

where E_d is the displacement energy, usually about 25 eV for metals. The volumetric source term of secondary atoms is thus

$$R' = \nu R = \frac{\Sigma_s \Phi \overline{E}_p}{2E_d} \tag{9.6}$$

However, these secondary atoms have a much smaller range $\overline{B}$ in the material. Again, for simplicity, assuming straight line paths, the number of atoms leaving the surface is

$$S' = R'\overline{B}/4 \tag{9.7}$$

and the sputtering ratio is

$$\Delta' = \frac{2S'}{\Phi} = \frac{R'\overline{B}}{2\Phi} = \frac{\Sigma_s \overline{E}_p \overline{B}}{4E_d} \tag{9.8}$$

Accounting for both the "collided" and "uncollided" ejected atoms, the total sputtering ratio Δ_T can be written approximately as

$$\Delta_T = \Delta + \Delta' = \frac{\Sigma_s}{2} \left(B + \frac{\overline{B}\ \overline{E}_p}{2E_d}\right) \tag{9.9}$$

The first determination of Δ_T was made for a gold foil exposed to a fission reactor spectrum of fast neutrons.

The measured value was $(1.0 \pm 0.3) \times 10^{-4}$ [Reference (9-2)]. Later the measurements were repeated for iron, yielding a value of $(5.7 \pm 0.8) \times 10^{-3}$ [Reference (9-3)]. Even later work showed no significant trend of Δ_T with temperature for iron [Reference (9-4)]. Using estimated values of the ranges, the values of the sputtering ratio calculated from Equation (9.9) are 5.6×10^{-7} for gold and 3.2×10^{-4} for iron, both considerably less than the experimentally determined values. Generally, theoretical predictions of fast neutron sputtering underestimate the experimental determinations. This gives evidence that the atoms are released from the surface in "clumps," rather than singly, but there is some dispute on this point [Reference (9-5)]. This being the case, comparison with theory would indicate that the average "clump" of atoms contains on the order of 10-100 atoms.

Illustration 9.1 From the measured neutron sputtering ratio of 6×10^{-3} for iron for 2-MeV fast neutrons, calculate $\bar{B}$, the average range of atoms in this material. Assume $E_d = 25$ eV and one atom released at a time. The sputtering ratio is (assuming secondary atoms predominate)

$$\Delta_T \simeq \frac{\bar{B}\Sigma_s \bar{E}_p}{4E_d}$$

where

$$\bar{E}_p = \frac{2A}{(A+1)^2} E_n = \frac{2 \times 56}{(56+1)^2} \times 2 \times 10^6 \text{eV} = 6.90 \times 10^4 \text{eV}$$

$$\Sigma_s = N\sigma_s = \rho \frac{N_A}{A_w} \sigma_s = 7.87 \times \frac{(6.02 \times 10^{23})}{55.85}$$

$$\times 11.4 \times 10^{-24} = 0.97 \text{cm}^{-1}$$

Δ_T and E_d are given as 6×10^{-3} and 25 eV respectively.

Therefore solving for $\bar{B}$, gives $\bar{B} = \dfrac{4 \times 25 \times 6 \times 10^{-3}}{0.97 \times 6.89 \times 10^{4}}$ $= 8.98 \times 10^{-6}$ cm.

This illustrates that fast neutron sputtering (to within an order of magnitude or so) is indeed a surface phenomenon.

That the "engineering" sputtering ratio is relatively large (measured values) could have significance in several applications. One of the direct effects is the physical erosion of the metal surfaces irradiated by fast neutrons. The number of atoms escaping from surface due to fast neutron of flux Φ is $\Delta_T\Phi/2$. Hence the physical erosion rate, C, defined by

Erosion rate = number of atoms escaping from surface per cm^2 per sec/atomic density,

is

$$C = \frac{\Delta_T \Phi A_W}{2\rho N_A} \quad \text{cm/sec} \qquad (9.10)$$

where

A_W = atomic weight
N_A = Avogadro's number
ρ = density, g/cm^3

For example, in a fast neutron flux of 10^{16} $n/(cm^2)(sec)$ such as might be encountered in a fast breeder reactor, this sputtering ratio indicates a metal removal rate of 4.1 mils/yr, which could seriously affect the integrity of the fuel cladding. Further, such sputtering is on the order of the metal removal rate found at operating temperatures in out-of-pile data for the corrosion process for stainless steel in contact with a liquid metal coolant [Reference (9-6)]. It should be noted here that the sputtering ratio is linearly dependent on the fast flux and thus the metal removal rate at lower fluxes would be proportionally lower while chemical corrosion processes would be essentially unchanged by a change in fast flux. The relatively large magnitude of the iron sputtering ratio also suggests that plasma thermonuclear devices may experience a contamination problem due to sputtering from the walls of such devices.

Fast neutron sputtering also causes continual freshening of surface (removal of oxide layers, etc.). These protective layers often inhibit chemical processes such as corrosion. Therefore, a higher corrosion rate of cladding materials in a reactor core is expected than measured "out-of-pile." This is likely to be the major consequence of fast neutron sputtering to a fission reactor.

Illustration 9.2 Find the erosion rate for iron due to fast neutron sputtering in the peak fast flux of Zimmer core, 2.5×10^{14} n/(cm^2)(sec). Erosion rate =

$$\frac{\Delta_T \Phi A_w}{2\rho N_A} = \frac{6 \times 10^{-3} \times 2.5 \times 10^{14} \times 55.85}{2 \times 7.87 \times 6.02 \times 10^{23}}$$

$$= 8.84 \times 10^{-12} \text{cm/sec}$$

$$\text{or} \quad = 0.11 \text{ mils/year}$$

This illustration indicates that for most LWR nuclear plants, the erosion of cladding by direct neutron sputtering is minor, since clad thicknesses are of the order of tens of mils.

9.2 Secondary Electron Emission from Charged Nuclear Particles

The details of the processes of secondary electron emission upon bombardment of a surface by high energy ions are not fully understood. Early work by Kapitza [Reference (9-7)] suggested that the secondary electron emission from energetic-ion bombardment could be interpreted as thermoionic emission resulting from local heating of surface material in a small area around the point of impingement. Modern theories, such as those presented by Sternglass [Reference (9-8)] and by Ghosh and Khare [References (9-9) and (9-10)] propose that secondary emission by such energetic ions can be explained on the basis of formation of secondaries by ionization of the target material, followed by diffusion of the secondaries to the surface and their subsequent escape. It is generally accepted that this latter mechanism is the more probable, although elements of both types of theories can be useful in interpreting the experimental results. In the theory presented by Sternglass [Reference (9-8)] it is assumed that the secondary yield is proportional to the stopping power of the material (dE/dx). Considering the energy loss by high-speed ions through ionization of the target material and considering the formation, diffusion, and escape of secondary electrons, Sternglass related the yield to the properties of the incident ion and characteristics of the target material as

$$\Delta = kfZ^{2/3} \left(\frac{Z_i^2}{E'}\right) \ln \left(\frac{4E'}{\bar{I}}\right) \qquad (9.11)$$

where k is a constant essentially independent of the target material, f is a factor slightly larger than unity, which accounts for the contribution to the yield from secondaries formed by the high-energy knock-on electrons (δ-rays) and can be approximated as $f = \frac{200 + E'}{100 + E'}$, Z is the atomic number of the target material, and $\bar{I}$ is the mean excitation potential for an atom of the stopping material. In this expression $E' = (m_e/M_i)\,E_i$ where m_e is the electronic mass and E_i, M_i, and Z_i are the energy, mass and charge on the incident ion. On the basis of experimental data, Sternglass evaluated k to be approximately 35.2 when

E' is expressed in electron volts. Equation (9.11) gives qualitative agreement with the observed secondary electron emission from metals bombarded by high-speed ions (ions with energy of the order of 100 KeV or larger).

Because of the insensitivity of the logarithmic factor, very crudely, Equation (9.11) suggests a dependency of the secondary electron yield with ion properties of the form

$$\Delta \propto \frac{Z_i^2 M_i}{E_i} \propto \left(\frac{Z_i}{v_i}\right)^2 \qquad (9.12)$$

where v_i is the velocity of the incident ion. For MeV-type protons, deuterons and alpha particles and for fission fragments at birth, v_i is of the same order of magnitude. Thus, the secondary electron yield from similar materials should progress as the square of the charge. Table 9.1 presents data indicating this trend. Note the striking effect of Z_i^2 for fission fragments. These data were obtained for fission fragments emerging from a thin film of UO_2. For fission fragments impinging upon UO_2, the value is roughly halved [Reference (9-15)].

9.3 Gamma-Induced Electron Emission

Another significant surface effect of radiation is the gamma-induced electron emission. This effect is particularly pertinent to the operation of electrical components in a radiation environment, and has been treated in detail in Chapter 5, Section 5.1, of this text.

9.4 Gamma Radiation-Induced Outgassing (GRIO)

As indicated in Chapter 5, Section 5.1, considerable high-energy electron emission occurs from surfaces exposed to gamma radiation. It was demonstrated in the early 1960s that high-energy electrons could appreciably desorb atoms from surfaces [Reference (9-16)]. In hindsight,

TABLE 9.1

Measured Secondary Electron Yields for Various Nuclear Charged Particles

Incident Particle	Particle Energy	Particle Charge	Material	Yield	Reference
Proton	0.5 MeV	+1	Al	1.58	(9-11)
Proton	2.5 MeV	+1	Al	0.50	(9-11)
Proton	0.5 MeV	+1	Mo	1.30	(9-11)
Proton	2.5 MeV	+1	Mo	0.61	(9-11)
Deuteron	0.5 MeV	+1	Al	2.29	(9-11)
Deuteron	2.5 MeV	+1	Al	1.51	(9-11)
Deuteron	0.5 MeV	+1	Mo	1.23	(9-11)
Deuteron	2.5 MeV	+1	Mo	1.09	(9-11)
Alpha	~0.7 MeV	~+ 1.35	Au	10.1 ± 1.9	(9-12)
Alpha	~3.3 MeV	~+ 1.86	Au	9.21 ± 0.01	(9-13)
Fission Fragments	~83 MeV	~+ 21	UO_2	570	(9-14)

of course, these early observations can readily be put together to indicate that gamma radiation will outgas surfaces in vacuum. This observation, however, appears at least in some respects to have had a different origin.

Real interest and thrust in gamma radiation-induced outgassing (GRIO) came with the U.S. space program. In the 1950s and early 1960s many research reactors were built across the country. With the interest in the effect of radiation on systems in vacuum, these sources of radiation were used to test various gear (usually electronics) in a vacuum environment. As an example of the interesting observations that evolved, a rather large-scale irradiation in vacuum was performed at the Battelle Research Reactor on electronic components, electromechanical devices and various materials. During a four-month irradiation period vacuum levels in a 6-inch-diameter continuously-pumped test section improved measurably, as shown by the readings from two gauges monitoring the system, shown graphically in Figure 9.1 [Reference (9-17)]. It appeared from these data that reactor radiations could significantly improve the vacuum level in a system. With this motivation, a more controlled experiment was performed as follows [Reference (9-17)]. A small stainless steel chamber (one liter) was close coupled to a compact getter-ion pump (11 liters/sec). A Phillips-type, cold-cathode gauge was attached to the chamber. The entire unit was enclosed in a watertight aluminum box. The chamber was evacuated (without bakeout) to 1.2×10^{-7} torr. It was then positioned in a pure reactor gamma radiation field (reactor shutdown residual gamma radiation) of about 10 R/hr at a chamber temperature of 92°F. The vacuum level in the chamber increased by about a decade during the first two hours and then slowly dropped. After 22 hours of exposure, the system was removed from the radiation field. The vacuum level then decreased rapidly for about 5 hours and finally leveled out in the mid 10^{-9} torr range, as shown in Figure 9.2. Thus the radiation exposure induced an improvement of nearly two orders of magnitude over the initial vacuum level. This experiment clearly established the existence of a significant GRIO.

About the same time (perhaps earlier), the first quantitative measurement of GRIO was performed by C.O. Muehlhause, et al, at the National Bureau of Standards [Reference (9-18)]. Data were obtained on a well-cleaned and baked 7071-6063 aluminum system exposed in both the NBS 32-kilocurie ^{60}Co gamma irradiation facility and the NRL 1-MW reactor facility. These data indicated that "the phenomenon of radiation-induced outgassing was principallv due to the gamma and its associated electron field rather than to the fast neutron field. The magnitude of the effect observed was $(7 \pm 4) \times 10^{-11}$ torr-liters/(sec)(cm^2) per watt/gm." [Reference (9-19)]

The second thrust in GRIO work came with the increased interest in the magnetically-confined fusion program. The primary concerns were the effects of radiation-induced outgassing from the first wall of the reaction vessel on vacuum level and on introduction of contaminants into the plasma. A summary of the rather sparse data on gas release from surfaces under X-ray impact up to 1975 is given by Brumbach and Kaminsky [Reference (9-19)]. Principal conclusions include:

"1. The dominant gas species released are O_2 and CO_2 and not H_2, H_2O or CO, which are typically observed in thermal desorption.

2. The mean quantum yields for O_2 and CO_2 release are in the range 8×10^{-5} to 2×10^{-3} molecules per photon in the tungsten bremsstrahlung spectrum at 50 KeV electron energy."

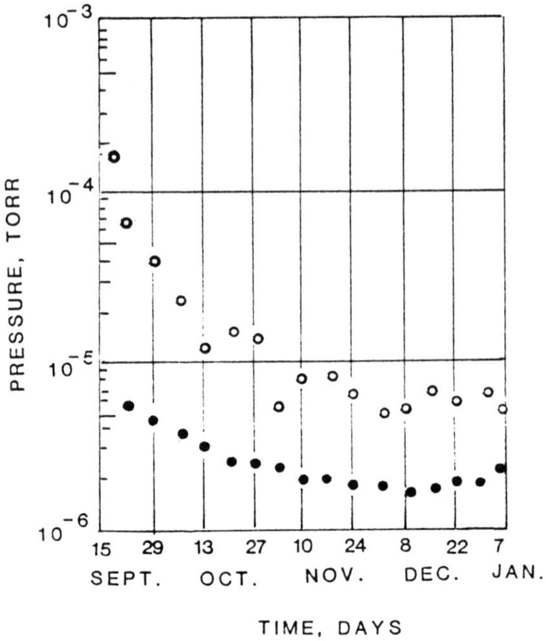

FIGURE 9.1 Vacuum levels during irradiation of experiment A.

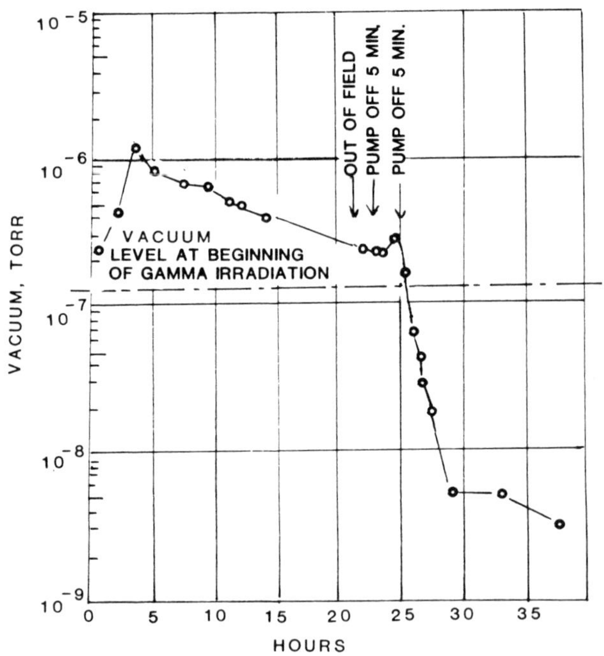

FIGURE 9.2 Vacuum levels in gamma field of 10^5 R/hr.

In 1976 Dobrozemsky [Reference (9-20)] reported observations on radiation-induced desorption from reactor radiations in which he concludes:

"1. For the complex structure of the beam-tube insert (mainly aluminum and copper, with small pieces of stainless steel and lead), the RID yields were found to be between 9 and 12 x 10^{-9} Pa $l/cm^2 s$ per Mrad/h (initial value) and 1 and 2 x 10^{-9} Pa $l/cm^2 s$ per Mrad/h (after 350,000-Mrad adsorbed dose).

2. RID is largely a bulk effect and does not depend a great deal on surface conditions."

In 1978 researchers at the University of Cincinnati began a systematic study of the effect of the base material on GRIO from ^{60}Co radiations. Using an ion-pumped 304 stainless steel vacuum system with test sections composed of tubes (or pins) of the various materials studied, results were obtained as indicated in Table 9.2.

In developing an analytical model for gamma radiation-induced outgassing, it is assumed that the electrons produced in the material by the various gamma ray interactions (Compton effect, photoelectric effect, and pair production) desorb atoms from the surface. Specifically, the model considers the following events:

1. Gamma rays produce electrons of average energy $\bar{E}_e$ within the material. An isotropic distribution of electrons is assumed.

2. The electrons "born" within a mean range $\bar{r}$ of the surface travel to the surface, losing energy in the process.

3. The primary electrons with reduced energy $\overline{E}_{\beta}(o)$ reach the surface and produce secondary electrons of average energy, $\overline{E}_s$.

4. The primary and possibly the secondary electrons cause electron-stimulated desorption of gas molecules from the surface.

The process of desorption outlined in these four steps is shown in Figure 9.3.

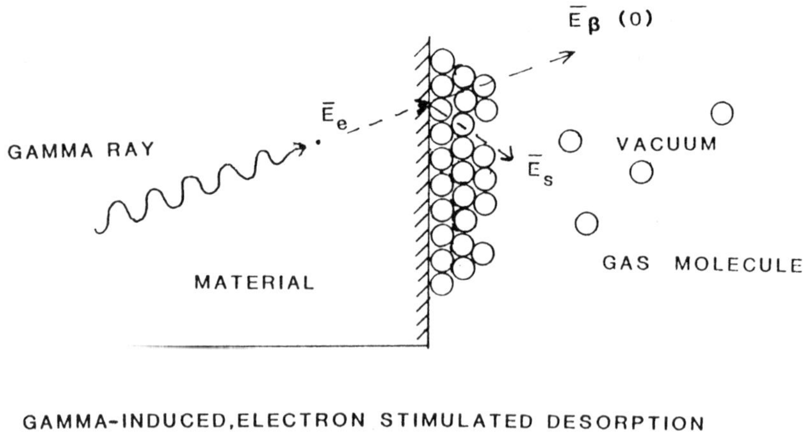

FIGURE 9.3 Gamma-induced electron stimulated desorption.

The interaction rate of gamma radiation with material is given as $\Sigma\Phi$ (primary gamma interactions)/(cm^3)(sec), where Σ is the macroscopic interaction cross section and Φ is the photon flux. Each primary gamma interaction produces a total of $\overline{\nu}$ free electrons, so that the volumetric source S of free electrons is

$$S = \Sigma\Phi\bar{\nu}\ \frac{(\text{electrons})}{(\text{cm}^3)(\text{sec})} \tag{9.13}$$

Approximately one fourth of these electrons from within a mean range $\bar{r}$ of the surface escape, so that the flux of electrons crossing the surface is*

$$\Phi_e = \frac{1}{4}\ S\bar{r} = \frac{\bar{r}\Sigma\Phi\bar{\nu}}{4}\ \frac{(\text{electrons})}{(\text{cm}^2)(\text{sec})} \tag{9.14}$$

These electrons desorb surface atoms with a yield Δ atoms per electron. Hence the release of atoms which have adsorbed to the surface is

$$\Phi_a = \frac{\bar{r}\Sigma\Phi\bar{\nu}\Delta}{4}\ \frac{(\text{atoms or molecules})}{(\text{cm}^2)(\text{sec})} \tag{9.15}$$

The usual expression of outgassing is in units of throughput (Pa-liters/sec) which is related to the atom flux by the ideal gas law on a per unit flux basis as

$$(PV) = \frac{\Phi_a RT}{N_A}\ \frac{(\text{Pa-liters})}{(\text{cm}^2)(\text{sec})} \tag{9.16}$$

where R is the gas constant, T is the absolute temperature and N_A is Avogadro's number. Substituting Equation (9.15) into Equation (9.16) gives

$$(PV) = \frac{\bar{r}\Sigma\Phi\bar{\nu}RT\Delta}{4N_A} \tag{9.17}$$

The absorbed radiation dose rate in the material $\dot{D}$ is related to the gamma flux as

---

* One half of the electrons formed are directed outward toward the surface. Of these, only one half are within the proper cone of emission, which is bounded by the mean range r. This expression neglects a (presumably) small contribution to the electron current from secondary electrons released from the surface by the primary electrons.

$$\dot{D} = \frac{\Phi \mu_a E_\gamma}{\rho} \tag{9.18}$$

where μ_a is the linear absorption coefficient of the material, ρ its physical density, and E_γ is the initial energy of the gamma radiation. Thus the theoretical expression for the gamma radiation-induced outgassing (GRIO) is given as

$$\text{GRIO} = \frac{(PV)}{\dot{D}} = (\rho\bar{r})\left(\frac{\Sigma}{\mu_a}\right)\frac{\bar{\nu}RT\Delta}{4N_A\, E_\gamma}\ \frac{(\text{Pa-liters})}{(\text{cm}^2)(\text{sec})}\ \text{per}\ \frac{(\text{megarad})}{(\text{hr})} \tag{9.19}$$

where appropriate conversion is required on the dose rate units. Σ is simply the linear attenuation coefficient μ. Inserting numerical values of $R = 8.29 \times 10^3$ (Pa-liters)/(gm-mole)(°K) and $N_A = 6.02 \times 10^{23}$ (particles)/(gm-mole), and with appropriate conversion factors, Equation (9.19) reduces to

$$(\text{GRIO}) = 5.96 \times 10^{-11}\ (\rho\bar{r})\ \left(\frac{\mu}{\mu_a}\right)\frac{\bar{\nu}T\Delta}{E_\gamma}\ \frac{(\text{Pa-liters})}{(\text{cm}^2)(\text{sec})\left(\frac{\text{megarad}}{\text{hr}}\right)} \tag{9.20}$$

where (ρr) is expressed in units of (gm)/(cm^2), T in Kelvin degrees, and E_γ in MeV. Equation (9.20) is the fundamental theoretical expression for determining the GRIO by this model.

The calculation of the high-energy electron release from various surfaces was presented in Sections 4.6 and 5.1, and is summarized in Table 5.3. The problem of determining the electron desorption yield Δ was considered best left to experimental evidence. Unfortunately, the available data in the electron energy range of interest are sparse. A summary of earlier data is given by Clausing [Reference (9-16)]. A portion of the experimental results are reproduced in Figure 9.4. These particular data were obtained for "large" vacuum systems for a copper surface first contaminated with diffusion pump oil and then after a mild thermal bake (20 hr. at temperatures from 100 to 145°C). It is seen that the surface cleanliness is extremely important in determining the value of Δ. In fact, for highly cleaned surfaces the values decrease by further orders of magnitude [Reference (9-24)]. The data shown in Figure 9.4 were selected to represent "typical" surfaces not highly conditioned, a situation corresponding to the majority of the GRIO experiments performed to date. From the data on Δ presented by Clausing, et al, [References (9-16) and (9-24)] concludes:

"1. the yield values are relatively large,

2. the yields decreased with bombardment time indicating a depletion of gas-producing material as the result of bombardment,

3. cleaning or thermal baking also reduced the yields, and

4. the yields per electron were independent of current at a given voltage so long as heating was avoided."

In most of the experiments reported in Table 9.2, a mild cleaning of the surfaces was performed as follows: acid etch, a distilled water rinse, followed by a rinse with ethyl alcohol. Thus it is believed that the corresponding Δ value would more appropriately be the lower curve of Figure 9.4, i.e., $\Delta \approx 0.1$.

With the rather large uncertainty in the desorption yield value of $\Delta \sim 0.1$, Table 9.3 presents a comparison of selected data from Table 9.2. Considering the uncertainties in both the experiment results and in the value of Δ, good agreement may be assumed. Note that the various differing factors in the theory [Equation (9.20)] all combine to produce an essentially constant predicted GRIO. That is, roughly speaking, all materials yield the same GRIO for cobalt-60 radiations, namely 10^{-9} (Pa-liters)/(cm^2)(sec) per megarad/hr. The marginal discrepancy of the carbon data with theory may possibly be attributed to lack of knowledge of the true or effective surface area for desorption.

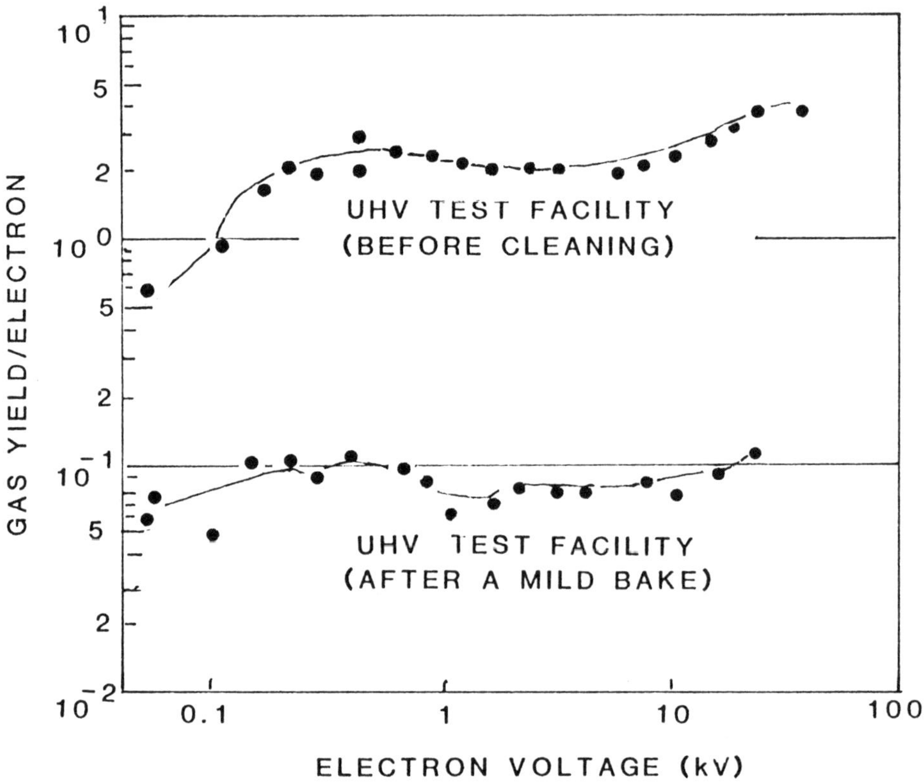

FIGURE 9.4 Yield of atoms desorbed per electron, Δ, versus electron energy.

[Reproduced with permission from Report ORNL-3652, Oak Ridge National Laboratory, operated by Union Carbide Corporation for the U.S. Department of Energy]

TABLE 9.2

Summary of GRIO Experiments at the University of Cincinnati (Cobalt-60 Gamma Irradiations)

Material	Surface-to-Volume Ratio, cm^{-1}	Pin Array	Investigator	GRIO Pa-liters/(cm^2)(sec) per megarad/hr
SS-304	8.64	Square	Toepker Ref. (9-21)	$(1.03 \pm 0.58) \times 10^{-9}$*
SS-304	8.64	Square	Shepherd Ref. (9-22)	$(1.42 \pm 1.75) \times 10^{-9}$
SS-304	5.70	Square	Shepherd Ref. (9-22)	$(1.16 \pm 0.66) \times 10^{-9}$
SS-304	3.08	Square	Shepherd Ref. (9-22)	$(0.96 \pm 0.38) \times 10^{-9}$
SS-304	8.64	Hexagonal	Shepherd Ref. (9-22)	$(1.33 \pm 0.41) \times 10^{-9}$
SS-316	7.75	Square	Shepherd Ref. (9-22)	$(1.18 \pm 0.49) \times 10^{-9}$
Aluminum	7.84	Square	Shepherd Ref. (9-22)	$(0.62 \pm 1.72) \times 10^{-9}$
Carbon	5.12	Square	Shepherd Ref. (9-22)	$(2.28 \pm 0.59) \times 10^{-9}$
Sorption Pump Sieve Material	3.37×10^6	Random	Patel Ref. (9-23)	$(0.70 \pm 0.22) \times 10^{-9}$

* Corresponds to a yield of 4.5×10^{-4} molecules per incident gamma-ray proton.

TABLE 9.3

Comparison of Theoretical Predictions with Selected Data

Material	GRIO, 10^{-9} (Pa-liters)/(cm^2)(sec) per ($\frac{megarad}{hr}$): Experiment	Theory[a]
SS-304	1.03 ± 0.58[21]	0.97
SS-304	1.42 ± 1.75[22]	0.97
Al	0.62 ± 1.72[22]	0.91
C	2.28 ± 0.59[22]	0.90
Sorption Pump Material[b]	0.70 ± 0.22[23]	0.90

[a] Assumes $\Delta = 0.1$. Average value of 1.25 MeV used for cobalt-60 gamma rays. Reference temperature of 293°K used.

[b] Z = 6.3.

REFERENCES

(9-1) Kaminsky, Manfred, Atomic and Ionic Impact Phenomena on Metal Surfaces, Academic Press, Inc., New York, 1965.

(9-2) Norcross, D.W., Fairand, B.P., and Anno, J.N., "Sputtering of Polycrystalline Gold by Fast Neutrons," J. Appl. Phys., 37, 621-623, 1966.

(9-3) Baer, T.S., and Anno, J.N., "Sputtering of Iron by Fast Neutrons," J. Appl. Phys., 43, 2453-2454, 1972

(9-4) Baer, T.S., and Anno, J.N., "Fast-Neutron Sputtering of Iron at Various Temperatures," J. Nuc. Matl., 54, 79-84, 1974.

(9-5) Harling, O.K., Thomas, M.T., Brodzinski, R.L., and Rancitelli, L.A., "Sputtering of Nb and Au by (D-T) Neutrons," Phys. Rev. Letters 34, No. 21, May, 1975

(9-6) Anno, J.N., and Walowit, J.A., "Analysis of Corrosion of Stainless Steel in a Sodium and High Radiation Environment," Jour. Nuc. Tech., 10, 67-75, Janaury, 1971.

(9-7) Kapitza, P.L., "On the Theory of -Radiation," Phil. Mag., 45, 989-998, 1923.

(9-8) Sternglass, E.J., "Theory of Secondary Electron Emission by High-Speed Ions," Phys. Rev., 108, 1-12, 1957.

(9-9) Ghosh, S.N., and Khare, S.P., "Secondary Electron Emissions from Metal Surfaces by High-Energy Ion and Neutral Atom Bombardments," Phys. Rev., 125, 1254-1258, 1962.

(9-10) Ghosh, S.N., and Khare, S.P., "Secondary Electron Emission from Metal Surfaces by H^+, H°, He^+, and He° Bombardments," Phys. Rev., 129, 1638-1642, 1963.

(9-11) Thornton, T.A., and Anno, J.N., "Secondary Electron Emission from 0.5-2.5 MeV Protons and Deuterons," Jour. Appl. Phys., 48, 1718-1719, 1977.

(9-12) Yarger, L.L., and Anno, J.N., "Secondary Electron Production from Approximately 1 MeV Alpha Particles Emerging from Gold," Jour. Appl. Phys., 37, 2929-2930, 1966.

(9-13) Anno, J.N., "Secondary Electron Production from Alpha Particles Emerging from Gold," Jour. Appl. Phys., 34, 3495-3499, 1963.

(9-14) Anno, J.N., "Secondary Electron Production from Fission Fragments Emerging from Thin Layers of Uranium Dioxide," Jour. Appl. Phys., 33, 1678-1681, 1962.

(9-15) Anno, J.N., and Jung, R.G., "Comparison of Secondary Electron Yield from Fission Fragments Emerging from and Impinging Upon a Thin Layer of UO ," Jour. Appl. Phys., 36, 3949, 1965.

(9-16) Clausing, R.E., "Release of Gases from Surfaces by Energetic Electrons," Oak Ridge National Laboratories, Oak Ridge, Tennessee, Report ORNL-Paper 818, 1964.

(9-17) Basham, S.J., Stang, J.H., "Experience with High Vacuum Irradiation Environments at the Battelle Research Reactor," Paper No. 41, International Symposium on In-Pile Irradiation Equipment and Techniques, Atomic Energy Research Establishment, Harwell, England, May 10-12, 1966.

(9-18) Muehlhause, C.O., Ganoczy, M., and Kupiec, C., "Radiation-Induced Outgassing in an Aluminium System," IEEE Transactions on Nuclear Science, February, 1965.

(9-19) Brumbach, S., and Kaminsky, M., "Gas Release from Surfaces Under X-ray Impact Photodesorption, Photocatalysis," Proceedings of the Symposium on Radiation Effects on Solid Surfaces, Meeting of the American Chemical Society, Chicago, Illinois, August 25, 1975.

(9-20) Dobrozemsky, R., "Operating Experience with an UHV System in the Vicinity of a Fission-Reactor Core," J. Vac. Sci. Tech., 19, 467-470, Jan/Feb, 1976.

(9-21) Toepker, T.P., and Anno, J.N., "Radiation-Induced Outgassing from Type 304 Stainless Steel," Nuc. Tech., 46, 127, September, 1979.

(9-22) Shepherd, A.R., and Anno, J.N., "Radiation-Induced Outgassing from Several Metals," Nuc. Tech., 52, 435, March, 1981.

(9-23) Patel, V.N., and Anno, J.N., "Radiation-Induced Outgassing from Sorption Pump Material," Nuc. Tech., 56, 401, February, 1982.

(9-24) ORNL-3652, Thermonuclear Division Semi-Annual Progress Report for Period Ending April 30, 1964, Oak Ridge National Laboratories, Oak Ridge, Tennessee, 132-143, July, 1964.

PROBLEMS FOR CHAPTER 9

1. Find the erosion rate for Zr cladding for fast flux 2.5×10^{14} n/(cm^2)(sec). Assume $\Delta_T = 6 \times 10^{-3}$.

2. An oxide layer of thickness of 100 Angstroms is formed on the zirconium cladding of a fuel element. This element is placed in the Zimmer core in an average fast neutron flux of 1.0×10^{14} n/(cm^2)(sec). Assuming no further oxidation, how long will it take for this oxide layer to be "eroded" away by fast neutron sputtering? Assume a sputtering ratio of $\Delta_T = 6 \times 10^{-3}$ and for estimate purposes that zirconium oxide has the same properties as zirconium metal. 1 Angstrom = 10^{-8} cm.

3. A stainless steel vacuum system contains 1000 cm^2 of surface area and is exposed at room temperature to an average gamma flux of 10^{13} photons/(cm^2)(sec). How many molecules/sec are released from the surface by the radiation?

10 MISCELLANEOUS RADIATION EFFECTS

In the earlier days of the atomic age, the effects of radiation were generally categorized under "radiation damage." However, with the passing of time, it became apparent that there were also many beneficial effects of radiation. The most familiar one, of course, is the use of radiation in cancer treatment. Considerable work has also been done, and is currently receiving revised interest in the area of food processing and preservation. This present discussion is primarily speculative on selected topics on other potential beneficial radiation effects.

10.1 Welding Plastics

It has been proposed that plastics can be welded by a neutron capture process [Reference (10-1)]. Preliminary experiments had already shown that plastics could be welded by unique processes involving thermal-neutron reactions [Reference (10-2)]. The surfaces to be welded are impregnated with boron- or lithium-containing material and exposed to thermal neutrons. The large cross-section reactions produce alpha particles by the following reactions:

$$^{10}_{5}B + ^{1}_{0}n \rightarrow ^{4}_{2}He + ^{7}_{3}Li + 2.79\ MeV$$

$$^{6}_{3}Li + ^{1}_{0}n \rightarrow ^{4}_{2}He + ^{3}_{1}H + 4.78\ MeV$$

$$^{7}_{3}Li + ^{1}_{0}n \rightarrow 2^{4}_{2}He + ^{0}_{-1}\beta + 16.2\ MeV$$

Some of the alpha particles (and recoil atoms) cross the interface, locally "stirring up" the plastic molecules and causing a local tack weld. After sufficient irradiation, the surface is mapped by the local welds, resulting in a bond.

The basic damage mechanism to organic or molecular materials resulting from bombardment by charged particles is the excitation of the orbital electrons. Displacement of atoms, which requires an energy transfer of ~25 eV, is relatively infrequent. An alpha-particle incident on an organic material will excite the orbital electrons of atoms within the molecules, either ionizing the atoms if sufficient energy is transferred or changing the electronic energy states to higher levels. The excited electrons will transfer their energy to the material lattice, resulting in broken molecular bonds, free radicals, and localized heating. The volume that is heated can be viewed as a cylindrical microvolume surrounding the alpha-particle track and the process can be treated with the thermal spike model. The temperatures are sufficiently high to melt the material in these microvolumes (e.g., it requires only 0.14 eV of dissipated energy to melt Mylar in the microvolume), and atoms or molecules will diffuse in these volumes during the period of time that the temperature spike endures. Since heating alone will not cause welding, it is assumed that the alpha particle, by momentum transfer or otherwise, mixes the molten material across the interface (i.e., accelerated diffusion occurs) so that on cooling, part of one material is actually imbedded into the other across the boundary.

From Equation (8.32), the cross-sectional area of

interface area which is heated above the melting point is

$$A = \left[\frac{dE}{dx}\right] / (ec\rho T_m) \qquad (10.1)$$

Considering an isotropic distribution of alpha particles at the interface results in an increase in the average area over normal impingement by a factor of $\pi^2/6$, then

$$\bar{A} = \pi^2 \left[\frac{dE}{dx}\right] / (6ec\rho T_m) \qquad (10.2)$$

This area is taken as the area of the local tack weld. Now the process of mapping the interface with "micro-areas" A is a statistical one, obeying the Poisson distribution. The probability of an area $\bar{A}$ not being mapped is $P_0 = \exp(-\delta t)$, where δ is the number of events per unit time. Considering the fact that the recoil atom will also create thermal spikes, δ is given as

$$\delta = \frac{N\sigma\Phi\pi^2 Q}{24ec\rho T_m} \qquad (10.3)$$

where

N = atomic density of the ^{10}B, ^{6}Li, or ^{7}Li

σ = microscopic thermal neutron reaction cross section

Φ = thermal neutron flux

Q = total energy release per neutron reaction

We now assume that the bond strength S is proportional to the fraction of the area welded and to the fraction F of the surface area effectively covered with the boron- or lithium-bonding agent. Then, for this simple theory

$$S = S_\infty F[1 - \exp(-\lambda\Phi t)] \qquad (10.4)$$

where

$\lambda = \delta/\Phi$

In some preliminary experiments testing the welding hypothesis, boron-10 in the form of boric oxide (B_2O_3) was selected as the neutron capture agent. The solvent or carrier of the boric oxide was a 20% solution of methyl alcohol (CH_3OH). Polymethylmethacrylate (PMMA) specimens were machined from 1/8-in.-thick (0.31 cm) strips beveled at 10 deg. Two such specimens were placed together to form a flat with contact surface area of ~0.5 in.^{2} (3.2 cm^2). The boric oxide solution was applied to the interface, the specimens were then clamped together and immediately irradiated at the core face of the Battelle Research Reactor. For successful welds, it was found that it was necessary to irradiate while the interface was wet with methyl alcohol and to apply pressure to the specimen to force better contact at the interface. Presumably by wetting the surface, the B_2O_3 became more uniformly distributed. After irradiation, the specimens were pulled apart, and the force required to break the bond was measured. Specimens without boron were prepared and irradiated under identical circumstances (including wetting the methyl alcohol) to ascertain that it was indeed the neutron capture process causing the welding.

For the first series of experiments, natural boron was used (19.8% ^{10}B). The PMMA specimens were irradiated to thermal-neutron fluences of 5×10^{13}, 10^{14}, 5×10^{14}, 10^{15}, 5×10^{15}, and 10^{16} n/cm^2 in groups of three. The maximum strength of each group was then chosen as the bond strength. The results are shown graphically in the lower curve of Figure 10.1. In general the bonding was very weak, with the maximum strength only ~20 psi at 10^{15} n/cm^2. It was observed that the specimens irradiated to fluences $>10^{15}$ n/cm^2 were very brittle, presumably due to damage by the fast neutrons and gamma rays present in the mixed radiation field at the face of the reactor. It is believed that such radiation damage is the cause of the decrease in bond strength above 10^{15} n/cm^2. In an effort to achieve larger bond strengths, two irradiations were performed using boron enriched to 94% in the ^{10}B isotope. As shown in the upper (dashed curve of Figure 10.1), a slight improvement was obtained.

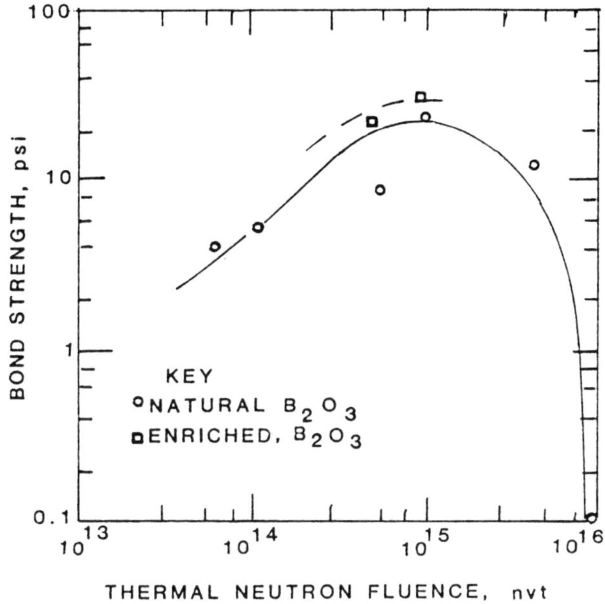

FIGURE 10.1 Bond strength versus thermal neutron fluence. [Reproduced with permission from J.N. Anno, "Welding Plastics by the Neutron Capture Process," Nuclear Technology, 29, pp. 124-126, 1976]

A least-squares fit of the data for the natural boron was made to the theoretical model [Equation (10.4)] for fluences up through 10^{15} n/cm^2, with the results $(S_\infty F)$ = 14.9 psi and $\lambda = 3.86 \times 10^{-15}$ per n/cm^2. With this value of λ, the effective ^{10}B concentration can be determined. Using the values Q = 2.79 MeV, σ = 3810 b, c = 0.35 cal/(g)(°C), ρ = 1.18 g/cm^3, and T_m = 180°C above room temperature, one obtains $N = 4.7 \times 10^{21}$ boron-10 atoms/cm^3. This compares favorably to the value of 6.3×10^{21} atoms/cm^3 for a completely saturated surface of B_2O_3, that is, for a complete layer of B_2O_3 deposited on the surface. Therefore, one concludes that in those areas where the boron compound is deposited, one is obtaining nearly as much boron which is theoretically possible. The problem with such weak welds would then appear to be that the fraction F of the surface that is welded is very small. In addition, the unfissioned B_2O_3 introduces foreign material between the surfaces, tending to weaken the bond.

Another extremely interesting observation from this analysis is to predict the absolute minimum neutron fluence to give a weld over 95% of the area, assuming F = 1 can be obtained. The most dense layer of ^{10}B one could deposit would be a thin layer of metallic boron, fully enriched, assuming that a density of 2.0 g/cm^3 gives a ^{10}B concentration of 1.2 x 10^{23} atom/cm^3. From the data fit and theory, this would result in a value of λ = 1.10 x 10^{-13} per n/cm^2. Thus, for $[1 - \exp(-\lambda\Phi t)] = 0.95$, we have $(\Phi t)_{minimum} = 3 \times 10^{13}$ n/cm^2. This value is within the realm of commercially available neutron generators, the use of which would also eliminate the fast-neutron and gamma-ray radiation damage to the plastic.

10.2 Gamma Sterilization of Plastic Medical Supplies

Gamma radiation doses of the order of megarads produce a high degree of kill of bacteria and viruses. For example, in water a dose of 10^5 rads will produce a 99.99 percent kill of microorganisms [References (10-3) and (10-4)]. It has been suggested that the use of gamma radiation, e.g., from cobalt-60, in appropriate dosages can sterilize polyethylene packages, which are common carriers of, themselves, plastic medical supplies. The primary problem with such usage is apparently the radiation damage to the polyethylene itself.

Figure 6.2 has shown the relative resistances of various plastics to radiation damage. It is seen that polyethylene exhibits intermediate damage behavior compared to other listed plastics. Damage in physical properties sets in at a gamma dose of about 10 megarads.

Detailed information on the effects of radiation on polyethylene are extracted from REIC Report No. 21 (Addendum) as follows [Reference (10-5)]:

"Polyethylene is unaffected by radiation to an absorbed radiation of 1.9×10^9 ergs g^{-1} (C)[a] and accrues 25 percent damage at 9.3×10^9 ergs g^{-1} (C). Tensile strength increases at first, but at approximately 1.1×10^{10} ergs g^{-1} (C), it begins to decrease, and is 25 percent lower than the initial value at approximately 10^{12} ergs g^{-1} (C)."

Polyethylene is subject to oxidation when irradiated. As a result, it is more stable in vacuum than in air.

Kerlin and Smith [Reference (10-6)] found that Marlex 6002, a high-density polyethylene, irradiated in air to an exposure of 10^9 ergs g^{-1} (C) decreased in elongation from 907 percent to 14 percent. However, in vacuum, the decrease was only to 675 percent. Tensile strength increased both in air and in vacuum but the increase was slightly higher in vacuum.

A common sterilization reagent currently in use for medical supplies is ethylene oxide. However, there are severe limits on the proper disposal of wastes of this reagent. It is therefore anticipated that the use of gamma radiation for this purpose will become more widespread in the future.

10.3 Improving Vacuum Levels with Gamma Radiation

As discussed in Chapter 9, Section 9.3, gamma radiation induces outgassing of surfaces in vacuum, and quite observable improvement results in vacuum systems as a result of exposure to gamma radiation. A question arises

---

[a] 1 rad = 100 ergs/gm.

as to whether gamma radiation-induced outgassing is a viable alternative to thermal bakeout. This alternative could be potentially useful for thermally-sensitive materials used in vacuum systems. To this end, a brief study was made with a sorption pump material [Reference (10-7)]. 345 g of molecular sieve material (as received, with no prior bakeout) was exposed in air at 20°C to ^{60}Co gamma radiation at a dose rate of 9.0×10^4 rads/hr, and the attendant outgassing measured. This outgassing was compared to thermal outgassing produced at various temperatures after the sieve material was exposed to a fixed quantity of air each time after initial bakeout at 300°C. By comparing the ultimate pressure reached in a closed container after bakeout at various temperatures, it was estimated that ^{60}Co GRIO at a dose rate of 9.0×10^4 rads/hr produced the same effect as a thermal bakeout at only 30°C. It was concluded that, at least for this particular material, GRIO is not a viable alternative to thermal bakeout [Reference (10-7)].

Although the above-reported results are not encouraging to the technique of GRIO for improving system vacuum levels, there might well be certain specialized applications in which such a process would be useful. Certainly the use of GRIO in ultra-ultra high vacuum studies may be worth exploration.

10.4 Use of GRIO to Measure Effective Catalyst Surface Area

Catalysts gained their considerable importance in the chemical industry from their ability to increase the rate and selectivity of chemical reactions. The catalyst has the function of lowering the activation energy barrier between reagents and reaction products. Numerous reactions between gaseous substances, of great importance to industry. are catalyzed by heterogeneous catalysts, the majority of which are solids and particularly metals [Reference (10-8)].

Without a catalyst, the reaction proceeds too slowly at temperatures favorable for thermodynamic equilibrium. At higher temperatures, the reaction rate is acceptable, but a shift of thermodynamic equilibrium will happen in a position in which conversion is restricted.

The great importance of measuring the specific surface area (area per unit mass) of a certain surface is due to the simple fact that the surface may simply represent a transient zone in exchange of material between the mass of the solid and the external environment. In catalysts, the effective surface area and porosity are the most important factors determining the activity, mechanism, and hence the selectivity of a catalyst.

A common method for determining the effective surface area is the gas adsorption (BET) method [Reference (10-9)]. However, this technique is rather tedious and time-consuming. It is suggested the gamma radiation-induced outgassing (GRIO) might be a more rapid and viable alternative to determine the effective catalyst surface area. Note from Chapter 9, Section 9.3, that the GRIO value is almost completely independent of the base material. One could thus determine the surface area by simply measuring the initial rate of change of pressure in a closed vessel containing the catalyst material exposed to gamma radiation. Such studies are being performed at the University of Cincinnati [Reference (10-10)].

10.5 Other Miscellaneous Radiation Effects

As the uses of radiation receive increased public acceptance, benefits similar to the ones suggested so far in this chapter might become more widespread. Numerous studies have been performed on its use to general public problems. As mentioned earlier in Section 10.2, sterilization of public water is a potential important use of gamma radiation. Treatment of sewage with radiation from spent fuel elements has also been explored [Reference (10-11)]. "Cold" sterilization of brewery products has also been suggested. Many of these thoughts, like food processing, require public acceptance, even after extensive research and testing.

REFERENCES

(10-1) Anno, J.N., "Welding Plastics by the Neutron Capture Process," Nuclear Technology, 29, pp. 124-126, 1976.

(10-2) Barkalov, J.M., Goldanskii, V.J., Gzantizev, B.G., and Yegorov, Y.V., High Molecular Compounds, 12, 1801, 1960.

(10-3) Research Report, "The Effects of Radiation on Chicago Metropolitan Sanitary District Municipal and Industrial Wastewater," Pacific Northwest Laboratories Division of Battelle Memorial Institute, Richland, Washington, March 12, 1968.

(10-4) Report, "The Practicality of Using Atomic Radiation for Wastewater Treatment," D.S. Ballantine and L.A. Miller, Division of Isotopes Development, AEC, and D.F. Bishop and F.A. Rhorman, Division of Research, Federal Water Pollution Control Administration, p. 20, no date.

(10-5) Radiation Effects Information Center Report REIC No. 21, Addendum, AD 454056, pp. 106-107, August 31, 1964.

(10-6) Kerlin, E.E., and Smith, E.T., "Measured Effects on the Various Combinations of Nuclear Radiation, Vacuum and Cryotemperatures on Engineering Materials," General Dynamics/Fort Worth, Fort Worth, Texas, FZK-172, September 30, 1963, Quarterly Progress Report, June 1-August 31, 1963, NAS8-2450, p. 121.

(10-7) Patel, Vikram N., and Anno, J.N., "Radiation-Induced Outgassing from Sorption Pump Material," Nuc. Tech., 56, 401, February, 1982.

(10-8) Gomer, R., Interactions on Metal Surfaces, Springer-Verlag, Berlin, 1975.

(10-9) Gregg, S.J., and Sing, K.S.W., Adsorption, Surface Area and Porosity, Academic Press, 1967.

(10-10) Allagi, M.O., Department of Chemical and Nuclear Engineering, University of Cincinnati, Cincinnati, Ohio, M.S. Thesis Research, 1983.

(10-11) Piccinni, Norberto, Simonetto, Rodolfo, and Del Tin, Giovanni, "Gamma Treatment of Sewage Sludge with Spent Nuclear Fuel," Nuc. Tech., 57, 243-253, May, 1982.

11 RADIATION EFFECTS ON BIOLOGICAL MATERIAL

Even though unknowns in the interaction between ionizing radiation and biological material still exist, more is known about radiation damage than any other etiologic agent in relation to cellular, molecular, and organ systems. Based upon this information, one is able to set proper exposure levels for the handling of ionizing agents, thereby limiting one's exposure rate.

11.1 Biological Damage

11.1.1 Dose-response. Over many years man has observed radiation effects on biological material and divided these effects into two major groups, threshold and non-threshold effects. A quantitative relationship is seen when one plots dose against effect as shown in Figure 11.1 [Reference (11-1)]. Curve (1) shows a threshold effect where a certain minimum dose must be exceeded before the effect under investigation is observed. Non-threshold effect is where no threshold is observed regardless of how small of a dose is applied [Curve (2), Figure 11.1].

In conveying information on radiation lethality (lethal dose rate), the field has accepted the terminology of the $LD_{50/30}$ value. This value relates the dose of radiation whereby 50% of the population will die within 30 days after exposure, i.e., it is the lethal dose to 50% in 30 days [References (11-2) and (11-3)]. Attempts have been made to measure or estimate the $LD_{50/30}$ in several species. Obtaining a value for this quantity depends upon dose rate, type of radiation, strain of species, cell cycle, and time of year, to mention a few. Values range from as high as 300,000 rads for paramecia to as low as 350 rads for dogs. But, in man the $LD_{50/30}$ value has been estimated to fall between 250 and 450 rads.

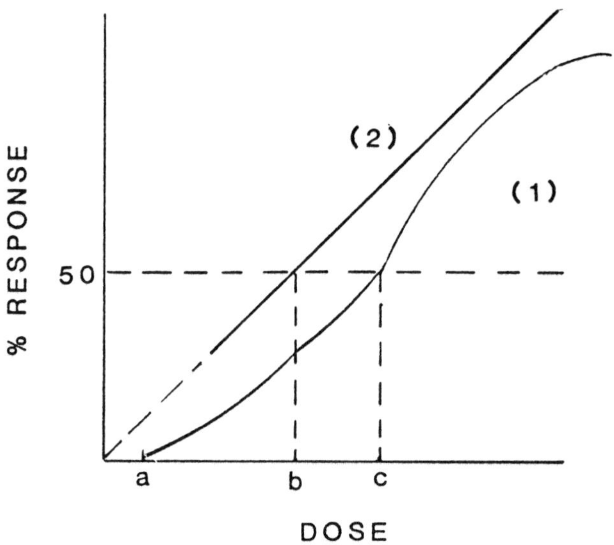

FIGURE 11.1 Dose-response curve. Curve (1) is threshold effect, with a as the threshold point and c as its 50% dose point. Curve (2) is non-threshold effect with b as its 50% dose point.

[Reproduced with permission from H. Cember, Introduction to Health Physics, Pergamon Press, 1969]

11.1.2 Direct and indirect ionizing interaction. Direct ionizing particles are electrically-charged particles having sufficient kinetic energy to produce ionization by collision. Indirectly ionizing particles are uncharged particles which can liberate directly ionizing particles. When the directly affected atom is in a protein molecule, or in a molecule of nucleic acid, the results can be a broken chain, causing molecule damage. However, the body is mostly made of water, so that an interaction causing energy to be deposited produces free radicals which are chemically toxic, where a free radical is a fragment of a compound or an element that contains an unpaired electron. When pure water is irradiated, an electron is ejected from a water molecule resulting in a positive ion and a negative electron, both which contain a great deal of energy [References (11-4) and (11-5)].

$$H_2O \rightarrow H_2O^+ + e^-$$

The electron is rapidly captured by another water molecule to yield a negative ion.

$$e^- + H_2O \rightarrow H_2O^-$$

The net result is two ions, positive and negative, which, in the presence of another water molecule, decompose into an ion and a free radical:

$$H_2O^+ \rightarrow H^+ + OH^\circ$$

and

$$H_2O^- \rightarrow OH^- + H^\circ$$

The ions H^+ and OH^- will recombine to form water, but the free radicals H° and OH° react with other molecules, causing toxicity. OH° radicals can combine with each other to produce hydrogen peroxide

$$OH + OH \rightarrow H_2O_2$$

while the H° radicals combine to form gaseous hydrogen.

The hydrogen peroxide, a powerful oxidizing agent, can then effect other non-damaged cells or molecules. Also, free hydrogen can combine with oxygen to form hydroperoxyl radicals

$$H + O_2 \rightarrow HO_2$$

which can combine with a free hydrogen radical to form hydrogen peroxide.

11.1.3 Radiation effects. An acute exposure is a single exposure to a high dose of radiation during a short period of time, and may cause biological effects in a short time after exposure. The effects vary from person to person and have been classified into three major categories [Reference (11-1)]:

1. Hemopoietic syndrome which appears due to a dose of 200 rads and is described as a depression or oblation of the bone marrow.

2. Gastrointestinal syndrome which occurs from a dose of 1000 rads or more and is characterized by desquamation of the intestinal epilhelium.

3. Central nervous system syndrome which appears after a dose of 2000 rads and causes central nervous system damage.

In all three cases, there are common effects of nausea and vomiting, malaise and fatigue, increased temperature, and blood changes. A summary of acute whole-body radiation effects can be found in Table 11.1 which covers dose received and the time associated with each effect [Reference (11-6)].

A chronic exposure is a radiation exposure of long

TABLE 11.1

Probable Early Effects of Acute Whole-Body Radiation Doses*

Acute Dose (rems)	Probable Observed Effect
5 to 75	Chromosomal aberrations and temporary depression of white blood cell levels in some individuals. No other observable effects.
75 to 200	Vomiting in 5 to 50% of exposed individuals within a few hours, with fatigue and loss of appetite. Moderate blood changes. Recovery within a few weeks for most symptoms.
200 to 600	For doses of 300 rems or more, all exposed individuals will exhibit vomiting within 2 hours or less. Severe blood changes, with hemorrhage and increased susceptibility to infection, particularly at the higher doses. Loss of hair after 2 weeks for doses over 300 rems. Recovery from 1 month to a year for most individuals at the lower end of the dose range; only 20% survive at the upper end of the range.
600 to 1000	Vomiting within 1 hour. Severe blood changes, hemorrhage, infection, and loss of hair. From 80 to 100% of exposed individuals will succumb within 2 months; those who survive will be convalescent over a long period.

* The whole-body doses given in his table are those measured in soft tissue near the body surface; because of energy absorption in the body, the interior (or vertical midline) doses, which are sometimes quoted. are about 70% of the values in the table.

[Taken from S. Glasstone and A. Sesonske, Nuclear Reactor Engineering, 3rd ed., D. Von Nostrand Company, New York, 1981]

duration delivered through fractionation. Greater lethality occurs due to radiation damage recovery of the short-time or acute exposure. If a certain amount of injury is required to produce death, and recovery starts as soon as the injury is produced, then some of the initial damage will repair before the lethal amount occurs. Among the consequences of accumulation of small doses is the development of cancer, life shortening, and cataracts.

Somatic mutations are mutations which take place in cells which are not germ cells. The mutations occur spontaneously and, as the radiation dose increases, the number of mutations increase. The effect is on an individual basis and the mutant genes are not passed on to offspring in sexual reproduction.

Gene mutations occur due to changes in the dexyribonucleic acid (DNA) structure and these changes involve either the base composition, the sequence, or both. Mutation production is influenced by many factors: cell type, dose rate, mitotic stage, sex, and species. Because of many variables, it is not easy to extrapolate known facts to predict the mutation rate in man. Information on radiation-induced mutation in man is lacking, even though studies have been conducted in relation to the Hiroshima and Nagasaki bombings. At this time, one is not able to detect mutations due to low level radiation, but there are estimates made in relation to "genetic deaths" which predict an increase of only 200 cases resulting from 1 rad above the total number of 240,000 which occur due to natural incidence [Reference (11-8)].

11.1.4 Natural and manmade radiation. There are many sources of natural background radiation and the dose value varies depending upon locality. Table 11.2 gives the average annual doses received by a person in the United States as of 1970 for natural and manmade radiation

TABLE 11.2

Sources of Natural and Manmade Radiation Sources and Accompanying Annual Doses as of 1970*

Natural Radiation		Dose, mrem/yr
External exposure sources		
Cosmic rays		45
Terrestrial γ-rays		60
Internal exposure sources		
^{40}K		17
Heavy elements		6.6
^{14}C		1.0
	Subtotal	129.6
Manmade Radiation		
Medical and dental X-rays		72
Radiopharmaceuticals		2
Global fallout		4
Nuclear power		0.003
Occupational		0.8
Miscellaneous		2
	Subtotal	80.8
Grand Total		210.4

* Based on Klement, et al, "Estimates of Ionizing Radiation Doses in the United States 1960-2000," U.S. Environmental Protection Agency Report ORP/CSD 72-1, 1972.

[Reproduced with permission from J.R. Lamarsh, Introduction to Nuclear Engineering, Addison-Wesley Publishing Company, Reading, Mass., 1975]

at sea level [Reference (11-7)].

As one moves to a higher altitude the exposure to cosmic radiation increases. For example, the exposure level of Denver, Colorado, is twice that of New York City and there is a 15% decrease at the equator compared to that at the poles. At 10,000 ft. the dose from cosmic rays increases to 90 mrems/yr and at 60,000 ft. the dose is 3 rems/yr for continuous whole-body exposure.

Terrestrial radiation will also vary from one location to another. A wooden house in New York City gives a dose of 45 mrems/yr whereas a house built in the southwest of rock will have a dose 15 to 20% higher. In India where one has thorium-bearing monazite sands, people are exposed to doses of 2 rems/yr for generations.

In the human body, one has natural occurring potassium-40 of about 130 grams for the average person, which produces an exposure of 17 mrems/yr. The body also contains carbon-14 and other heavy elements producing an exposure of 1 to 6.6 mrems/yr, respectively.

Manmade radiation varies in the amount of exposure it causes, from .003 mrems/yr for nuclear power to 72 mrems/yr for medical and dental X-rays. Global fallout due to bomb testing is estimated at 4 mrems/yr as of 1972, and this value is decreasing as the number of above ground tests are ceasing. Combining these with radiopharmaceuticals, occupational exposure, and other uses such as industry, one is receiving a total manmade exposure of 80.8 mrems/yr in the United States. This, combined with a natural exposure of 129.6 mrems/yr, gives a total average exposure of 210.4 mrem/yr for a person living in the United States. Note that this chronic yearly exposure is a factor of about 1000 less than the acute $LD_{50/30}$ exposure.

Although debates continue to rage on the effects of low-level radiation on man, there is indeed scientific evidence that some radiation is biologically beneficial [Reference (11-9)]. Further, it is interesting to note, for whatever value, that the highest natural background on earth is found in the Euphrates Valley, the Biblical cradle of mankind.

11.2 Biological Radiation Units

One of the most important components in the development of one's understanding of biological effects is a knowledge of the terminology involved in determining radiation effects. With this in mind, this section is devoted to defining the terms and units necessary for a general understanding.

As mentioned in Chapter 2, the Roentgen (R) is a unit of exposure, where exposure is a measure of the ionization produced in air by X- or gamma radiation. It is the sum of the electrical charges on all ions of one sign produced in air when all electrons liberated by photons in a volume element of air are completely stopped in air, divided by the mass of the air in the volume element. $1R = 2.58 \times 10^{-4}$ Coulomb/kg of air. The rad is a unit of absorbed dose equal to 100 ergs/gm or hence 0.01 J/kg (0.01 Gy), in any medium. The rem is a unit of dose equivalent which is numerically equal to the dose in rads multiplied by the appropriate modifying factor such as RBE or QF (see below). The advantages of the rem unit as a measure of biological effects is that the rem doses from various types of radiation (alpha, beta, gamma, etc.) are directly additive. This is particularly useful in keeping exposure histories of radiation workers.

The Relative Biological Effectiveness (RBE) is a factor used to compare the biological effectiveness of absorbed radiation doses due to different types of ionizing radiation. It is the ratio of the amount of energy of any radiation required to produce a given effect to the energy required of 200 KeV X-rays to produce the same effect. Table 11.3 gives a listing of the RBE for various radiations [Reference (11-10)].

For most "engineering" health physics purposes, the use of the rem as simply (rem = rad dose x RBE) is sufficient, and has been standard practice in the field. However, in recent years, a refinement to the RBE, termed the Quality Factor (QF), has been introduced. Instead of basing biological effects on a comparison basis to 200 KeV X-rays (the RBE), the Quality Factor bases the biological effects on the Linear Energy Transfer (LET). The LET is the rate at which charged particles impart energy locally to a medium and is a function of their energies and velocities. Table 11.4 gives the relationship between Linear Energy Transfer and Quality Factor [Reference (11-11)]. Based upon the LET, the Quality Factor for various radiations is tabulated in Table 11.5 [Reference (11-7)]. Thus more modern practice has the rem determined as (rem = rad dose x QF).

A modern unit in less common engineering use is the Kerma. The Kerma is the sum of the initial kinetic energies of all charged particles liberated by indirectly ionizing particles in a volume, divided by the mass of matter in that volume. It depends only on the amount of energy transferred from the incident radiation to the material at the point of collision.

The curie is a unit of activity which equals 3.700×10^{10} nuclear transformations per second. For most practical purposes, fractions of the curie unit are used:

TABLE 11.3

Recommended Values of Relative Biological Effectiveness

Radiation	RBE	Biological Effect
X-rays, gamma rays, beta particles and electrons of all energies	1	Whole body irradiation (blood-forming organs critical)
Fast neutrons and protons up to 10 MeV	10	Whole body irradiation (cataract formation critical)
Thermal neutrons	2.5	Whole body irradiation
Naturally occurring alpha particles	10	Carcinogenesis outside skeleton

[Taken from S. Glasstone and A. Sesonske, Nuclear Reactor Engineering, D. Von Nostrand Company, Princeton, NJ, 1963]

TABLE 11.4

LET - QF Relationship

LET_{∞} (KeV/μ in water)	QF
3.5 or less	1
3.5 - 7.0	1 - 2
7.0 - 23	2 - 5
23 - 53	5 - 10
53 - 175	10 - 20

[Reproduced with permission from "Report of the RBE Committee to the International Commissions on Radiological Protection and on Radiological Units and Measurements," Health Physics, 9, No. 4, 381, April, 1963]

TABLE 11.5

Quality Factors for Various Types of Radiations*

Type of Radiation	Q
X-rays and γ-rays	1
β-rays, $E_{max} > 0.03$ Mev	1
β-rays, $E_{max} < 0.03$ MeV	1.7
Naturally occurring α-particles	10
Heavy recoil nuclei	20
Neutrons:	
Thermal to 1 KeV	2
10 KeV	2.5
100 KeV	7.5
500 KeV	11
1 MeV	11
2.5 MeV	9
5 MeV	8
7 MeV	7
10 MeV	6.5
14 MeV	7.5
20 MeV	8
Energy not specified	10

* Based on NCRP Report 39.

[From Lamarsh, Ref. (11-7)]

Millicurie (mCi) = 3.7×10^7 disintegrations/sec
Microcurie (μCi) = 3.7×10^4 disintegrations/sec
Picocurie (pCi) = 3.7×10^{-2} disintegrations/sec

Two new SI units have been added to the terminology: the Gray and the Sievert. The Gray is a unit of absorbed dose equivalent to 100 rads and the Sievert is a unit of biological dose equivalent to 100 rems.

11.3 Radiation Protection Standards

There are several agencies which set standards or guidelines in radiation protection. The International Commission on Radiation Protection (ICRP) whose basic responsibility is providing guidance in matters of radiation safety, was established in 1928. Its main concern was initially with the safety aspects of medical radiology, but as time went on and the use of radioactive materials increased, the agency's interests expanded outward from medical radiology. ICRP official policy is in preparing recommendations on the basic principles of radiation protection and leaving to national regulations the responsibility of introducing the detail regulations.

In the United States, the National Committee on Radiation Protection and Measurements (NCRP) was formed in 1929 to undertake the ICRP recommendations on a national level. The NCRP is concerned only with the scientific and technical aspects of radiation protection and is constructed of technical experts in radiation protection and of scientists with disciplines directly related to the field of radiation protection.

A second national body was formed by the President of the United States in 1959, and incorporated into the Environmental Protection Agency (EPA) in 1970, this being the Federal Radiation Council (FRC). The purpose of the

FRC is to guide the government in the formulation of Federal policy on human radiation exposure. The FRC covers not only technical and scientific consideration, but also non-scientific aspects such as political, military, economic, and social considerations.

The U.S. Nuclear Regulatory Commission (NRC) regulations are based entirely upon the guidelines of the ICRP and the NCRP. It should be noted here that the NRC does not set radiation standards for the public at large, but rather establishes regulations in regard to permissible levels of radiation in nuclear power plants and plant site boundaries.

To establish limits of radiation exposure, a model is needed to be constructed. Since there is no evidence of deleterious effects of radiation at low doses, to handle the risk of low-level radiation exposure to humans, one must extrapolate from the well established effects of high-level exposures that have been developed through studies of the Japanese survivors of the atomic bombings, uranium miners, radium-dial painters, and radiologists. There are four trends of thought on the extrapolation to lower levels.

The most accepted rule is one of linear extrapolation, i.e., at low dose levels the risk of cancer is directly proportional to the dose (Curve a of Figure 11.2) [Reference (11-13)]. Other researchers adopt a threshold effect, below which no risk is incurred (Curve b). Still others contend that the risks are disproportionately lower than expected by linearity (Curve c) and still others take the view of risks being higher than that predicted by linearity (Curve d).

The viewpoint that has been generally adopted is one of a linear dose-effect relationship at lower dose levels.

There is no threshold effect in radiation dose, all doses delivered are additive, and there is no biological recovery at low doses. This, therefore, gives a conservative base upon which to establish standards [Reference (11-14)].

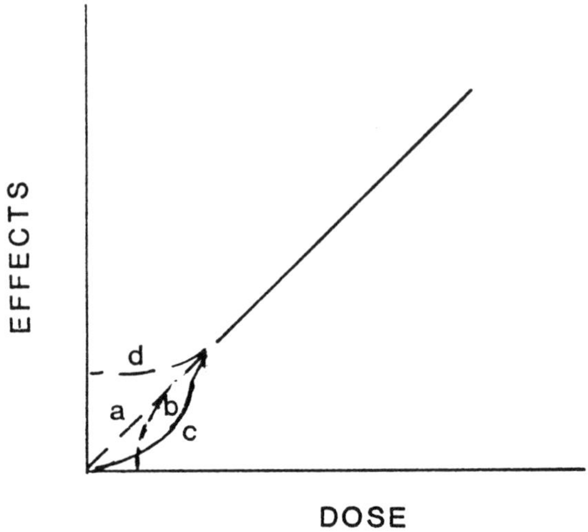

FIGURE 11.2 Dose-effects curve.

[Reproduced with permission from BEIR Report, "Biological Effects of Ionizing Radiation," National Academy of Sciences, Washington, D.C., 1980]

Both the ICRP and the NCRP give radiation standards in terms of maximum permissible dose (MPD), where ICRP defines MPD as:

"The permissible dose for an individual is that dose, accumulated over a long period of time or resulting from a single exposure, which in the light of present knowledge, carries a negligible probability of severe somatic or genetic injuries; furthermore, it is such a dose that any effects that ensued more frequently are limited to those of a minor nature that would not be considered unacceptable by the exposed individual and by competent medical authorities." [Reference (11-15)] Any severe somatic effects (e.g., leukemia) that might result from exposure of individuals to the permissible dose would be limited to an exceedingly small fraction of the exposed group; effects such as shortening of life span, which might be expected to occur more frequently, would be very slight and would likely be hidden by normal biological variations. The permissible doses can therefore be expected to produce effects that could be detected only by statistical methods applied to large groups.

Dose limiting recommendations of the NCRP can be found in Table 11.6 [Reference (11-7)]. Recommendations have been divided into two groups:

1. occupational exposure, i.e., adults whose course of work exposes them to ionizing radiation, and

2. the general public.

The prospective annual limit is 5 rems in any one year for occupational workers and one tenth of that for an individual of the general public. This does not include exposure due to medical X-rays or any other medical-related exposure. Also, the total accumulated dose a worker can receive to the age of N shall not exceed 5(N-18) rems. The recommendation of 5 rems in any one year is averaged over a 50-week period yielding a 100 mrems per week MPD.

After World War II a need was seen in the assessment of risk in relation to radiation exposure. To help in this, the National Academy of Sciences formed the Committee on the Biological Effects of Ionizing Radiation (BEIR). The studies by the BEIR committee brought out dose-effect models which were mentioned above in the development of MPDs.

The risk estimate for cancer is about 100 cases per million person-rem, i.e., if a million persons are exposed to one rem above natural background during their lifetime,

TABLE 11.6

Dose-Limiting Recommendations of the National Council on Radiation Protection and Measurements

Type of Exposure - Group	Maximum Permissible Dose Equivalent
Occupational exposure	
Whole body:	
prospective annual limit	5 rems, any one year
retrospective annual limit	10-15 rems, any one year
long-term acculumation to N years of age	5(N-18) rems
Skin	15 rems, any one year
Hands	75 rems, any one year (25/qtr)
Forearms	30 rems, any one year (10/qtr)
Other organs	15 rems, any one year (5/qtr)
Fertile women	0.5 rem in gestation period
Public or occasionally exposed individual	
Individual or occasional	0.5 rem, any one year
Students	0.1 rem, any one year
Population as a whole	0.170 rem, average per year
Emergency	
Lifesaving	
Whole body (preferably older than 45 years)	100 rems*
Hands, forearms	200 rems additional
Less urgent	
Whole body	25 rems
Hands, forearms	100 rems, total

* Preferably, once-in-a-lifetime exposure.

[From Lamarsh, Ref. (11-7)]

then the expected number of cancers in the group would be increased above the normally expected 200,000 cancer deaths by about 100. Such a small increase could not be detected by statistical means, given the normal variability in cancer frequency. Applying this estimate to the 60,000 people in the United States who work with radiation and assuming 40 years of continuous work at the average exposure of 0.6 rad/year which is experienced today, one obtains an increase, due to occupational exposure, of less than 3% [Reference (11-13)].

In recent years the BEIR committee has published a BEIR III report (in 1980) [Reference (11-15)] to update their earlier findings. One of the important changes is that the cancer risk from low-level, low linear-energy-transfer radiation was estimated using a linear-quadratic function to extrapolate to low doses from more reliable data at high dose levels. The upper and lower bound estimates were obtained using the linear model and the pure quadratic model, respectively. This model predicts that a single dose of 10 rads to 1 million persons would result in a 0.47% increase in cancer deaths over the previous model. For a continuous lifetime exposure to 1 rad/yr, the linear-quadratic model predicts an increase in cancer deaths of 2.8%. And for an exposure that simulates an occupational situation, 1 rad/yr from age 20 to 65, the linear-quadratic model predicts increase of 1.9% deaths. [For a more indepth coverage see References (11-12), (11-15), (11-16), (11-17), and (11-18)].

11.4 Exposure Calculations

11.4.1 External dose. In order to evaluate the hazard of radiation exposure, one must be able to calculate the exposure due to the presence of radioactive materials. The purpose here is not to give an indepth coverage of the subject, but to introduce one to the calculations which are used in obtaining an estimated exposure. The exposure rate due to an external source is obtained in the following way:

$$\text{exposure rate} = \frac{6CEf}{d^2}\ \text{mR/hr} \qquad (11.1)$$

where

C = source activity, mCi
E = particle energy, MeV
d = distance away from source, ft
f = the fraction of time the particle appears per disintegration

Illustration 11.1 A radiation worker is at a distance of 12 feet from an unshielded 0.3 Ci, Cs-137, point source which emits a 0.662 MeV gamma ray 100% of the time. What is the exposure rate at this distance?

Here we can apply Equation (11.1) directly where C = 300 mCi, E = 0.662 MeV, d = 12 ft., and f = 1

$$\text{exposure rate} = 6\ CEf/d^2$$

$$= \frac{6(300)(.662)(1)}{(12)^2}$$

$$= 8.275\ \text{mR/hr}$$

Consider the dose due to beta-particles. This can additionally be evaluated by use of a basic equation which incorporates the beta-particle flux, φ_β and the stopping power, S_p. The beta dose is given as

$$\dot{D} = .0576\ \varphi_\beta\ S_p\ \text{mrad/hr} \qquad (11.2)$$

External exposure to gamma rays is measured in Roentgens (see Chapter 2). The rate at which gamma-ray energy is deposited is determined by the gamma-ray intensity, its energy, and the mass absorption coefficient which is based on the gamma-ray's energy (see Table 4.5). Thus the exposure rate $\dot{X}$ is just

$$\dot{X} = 1.83 \times 10^{-8}\ IE(\mu_a/\rho)^{air}\ R/sec \qquad (11.3)$$

$$= 0.0659\ IE(\mu_a/\rho)^{air}\ mR/hr$$

where

I = gamma-ray intensity, (photons)/(cm^2)(sec)
E = gamma-ray energy (MeV)
$(\mu_a/\rho)^{air}$ = mass absorption coefficient of air at energy E

<u>Illustration 11.2</u> What is the exposure rate given a beam intensity of 400 gamma rays/(cm^2)(sec) of energy E = 1 MeV? From Table 4.5, $(\mu_a/\rho)^{air}$ = 0.0280 cm^2/g at 1 MeV. Applying this value in Equation (11.3) gives

$$\dot{X} = 0.0659\ (400)(1)(0.0280)$$
$$= 0.739\ mR/hr$$

The dose rate, i.e., the amount of energy deposited per unit time in tissue due to an external source of gamma rays is measured in rads. 1 rad, equivalent to the absorption of 100 ergs of gamma-ray energy per gram, is also equivalent to 6.25 x 10^7 MeV/g. As in the exposure to air, one has the absorption rate based on the gamma-ray intensity, energy, and mass absorption coefficient in tissue. Therefore, the absorbed dose rate $\dot{D}$ is given as

$$\dot{D} = 1.60 \times 10^{-8}\ IE(\mu_a/\rho)^{tiss}\ rad/sec \qquad (11.4)$$
$$= 0.0576\ IE(\mu_a/\rho)^{tiss}\ mrad/hr$$

where $(\mu_a/\rho)^{tiss}$ = mass absorption coefficient of tissue at energy E.

Illustration 11.3 Based upon Illustration 11.2, what is the rad dose rate? From Table 4.5 $(\mu_a/\rho)^{tiss} = 0.0300\,cm^2/g$ at 1 MeV. Applying this value to Equation (11.4) gives

$$\dot{D} = 0.0576\ (400)(1)(.0300)$$
$$= 0.691\ mrad/hr$$

The relationship between the exposure in air and the absorbed dose in tissue is the relation between the two equations

$$\dot{D} = \frac{0.0576 IE(\mu_a/\rho)^{tiss}}{0.0659 IE(\mu_a/\rho)^{air}}\ \dot{X}$$
$$= 0.874\ \frac{(\mu_a/\rho)^{tiss}}{(\mu_a/\rho)^{air}}\ \dot{X} \qquad (11.5)$$

To obtain the dose due to an external beam of neutrons one must consider the energy deposited to each of the tissue elements that react with the neutrons. Neutron relations are based on the neutron energy; for energies up to 20 MeV (fast neutrons) the reaction is one of elastic collisions. Here, the energy is dissipated in the immediate vicinity of the reaction. This interaction gives a collision dose rate of

$$\dot{D}_n(E) = \frac{\Phi E\ \sum_i N_i \sigma_i f_i}{100}\ \text{rad/sec} \qquad (11.6)$$

where

Φ = neutron flux, $n/(cm^2)(sec)$
E = neutron energy, ergs
N_i = atoms per gram of ith element
σ = microscopic scattering cross section of the ith element
f = mean fractional energy transfer
$= \frac{2M}{(M + 1)^2}$, where M = mass number of the element

Illustration 11.4 The fast neutron flux is 200 n/(cm^2)(sec) of energy E = 20 MeV at a particular point. What is the dose rate to soft tissue given the following?

Element	σ, cm^2
H	1.55×10^{-24}
C	1.65×10^{-24}
N	1.00×10^{-24}
O	1.55×10^{-24}

First consider the hydrogen element for Equation (11.6)

$$\dot{D}_n(H) = \frac{\Phi E N_H \sigma_H f_H}{100}$$

$$= (200\ n/cm^2 sec)(20\ MeV/n)(1.6 \times 10^{-6} erg/MeV) \times \frac{(6.7 \times 10^{22}\ atoms/g)(1.5 \times 10^{-24}\ cm^2/atoms)(.5)}{100\ ergs/g/rad}$$

$$= 3.22 \times 10^{-6}\ rads/sec$$

$$= 11.58\ mrads/hr$$

Based on the same equation we have

$\dot{D}_n$ (C) = 0.226 mrad/hr
$\dot{D}_n$ (N) = 0.033 mrad/hr
$\dot{D}_n$ (O) = 1.16 mrads/hr

Therefore the total fast neutron dose rate is the sum of the parts

$\dot{D}$ = 13.0 mrads/hr

For neutron energies in the thermal range, the thermal neutrons may be captured and may result in two reactions,

$^{14}N(n,p)^{14}C$ and $^{1}H(n,\gamma)^{2}H$. For the neutron-proton reaction, the dose rate can be calculated as

$$\dot{D} \doteq \frac{\Phi N \sigma Q (1.6 \times 10^{-6})}{100} \text{ rad/sec} \qquad (11.7)$$

where

Φ = thermal flux, n/(cm^2)(sec)
N = number of nitrogen atoms/g
= 1.17×10^{21}
σ = microscopic absorption cross section for nitrogen = 1.75×10^{-24} cm^2
Q = energy released by the reaction
= 0.63 MeV

Illustration 11.5 The thermal neutron flux at a particular point is 3000 n/(cm^2)(sec). What absorbed dose rate will one receive standing at that point? Equation (11.7) gives

$$\dot{D} = \frac{(3000)(1.17 \times 10^{21})(1.75 \times 10^{-24})(.63)(1.6 \times 10^{-6})}{100}$$

$$= 6.1 \times 10^{-8} \text{ rads/sec}$$

$$= 0.22 \text{ mrads/hr}$$

For the neutron-gamma reaction, the calculation for the dose rate is the same as that for a gamma-ray beam. [For a more indepth coverage see References (11-5) and (11-16).]

11.4.2 Internal dose. Charged particles having energies found in general nuclear engineering (mostly beta particles) have a short range in tissue. Thus the health hazard is not in external irradiation as much as it is in internal irradiation, which can become a very serious problem.

Knowing the concentration, C(t), that is, the number

of microcuries at a given time present in an organ of the body, one can obtain the dose with a sufficient accuracy. One assumes all of the energy is absorbed (due to the charged particles' short range) and recalling the 1 μCi = 3.7×10^4 disintegrations per sec one can find the dose rate as

$$\dot{D} = \frac{5.92 \times 10^{-4} C(t)\ E}{M} \text{ rad/sec} \qquad (11.8)$$

where

C(t) = concentration, μCi
E = average energy, MeV
M = mass of the organ of interest [Table 11.7, Reference (11-19)]

Knowing the quality factor, (QF) one can obtain the dose equivalent rate

$$H = \frac{5.92 \times 10^{-4} C(t)\ E(QF)}{M} \text{ rem/sec} \qquad (11.9)$$

Illustration 11.6 What is the internal dose rate equivalent due to a 15 mCi ingestion of tritiated water? From Tables 11.5 & 11.7, QF = 1 and M = 43,000 grams for body water. Therefore, applying these values gives, from Equation (11.9),

$$H = \frac{5.92 \times 10^{-4} (15000)(.006)(1)}{43000}$$
$$= 1.23 \times 10^{-7} \text{ rems/sec}$$
$$= 0.446 \text{ mrem/hr}$$

To obtain the amount of energy deposited over a time period, one must not only consider the radioactive decay of the radionuclide, but also the biological decay, which together give the "effective half-life." The biological decay is based upon the retention of the radionuclide in the organ. The effective half-life, T_e^h, is based on the

TABLE 11.7

Organs of Standard Man

(Mass and Effective Radius of Organs of the Adult Human Body)

	Mass, m (g)	Percent of Total Body	Effective Radius, X (cm)
Total Body	70,000	100	30
Muscle	30,000	43	30
Skin and subcutaneous tissue	6,100	8.7	0.1
Fat	10,000	14	20
Skeleton			
Without bone marrow	7,000	10	5
Red marrow	1,500	2.1	
Yellow marrow	1,500	2.1	
Blood	5,400	7.7	
Gastrointestinal tract	2,000	2.9	30
Contents of GI tract			
Lower large intestine	150		5
Stomach	250		10
Small intestine	1,100		30
Upper large intestine	135		5
Liver	1,700	2.4	10
Brain	1,500	2.1	15
Lungs (2)	1,000	1.4	10
Lymphoid tissue	700	1.0	
Kidneys (2)	300	0.43	7
Heart	300	0.43	7
Spleen	150	0.21	7
Urinary bladder	150	0.21	
Pancreas	70	0.10	5
Salivary glands (6)	50	0.071	
Testes (2)	40	0.057	3
Spinal cord	30	0.043	1
Eyes (2)	30	0.043	0.25
Thyroid gland	20	0.029	3
Teeth	20	0.029	
Prostate gland	20	0.029	3
Adrenal glands or suprarenal (2)	20	0.029	3
Thymus	10	0.014	
Ovaries (2)	8	0.011	3
Hypophysis (Pituitary)	0.6	8.6×10^{-4}	0.5
Pineal gland	0.2	2.9×10^{-4}	0.04
Parathyroids (4)	0.15	2.1×10^{-4}	0.06
Miscellaneous (blood vessels, cartilage, nerves, etc.)	390	0.56	

[Reprinted with permission from Y. Wang, Handbook of Radioactive Nuclides, (c) CRC Press, Inc., Boca Raton, FL, 1969] effective decay constant, λ_e, which equals ln2 divided by the physical plus biological decay constants

$$T_e^h = \frac{\ln 2}{\lambda_e} = \frac{\ln 2}{\lambda_p + \lambda_b} = \frac{1}{\frac{1}{T_p^h} + \frac{1}{T_b^h}} \qquad (11.10)$$

where T_p^h and T_b^h are physical and biological half-lives, respectively.

Recall that the fractional number of atoms disintegrating in a time interval t is $(1-e^{-\lambda_e t})$. The total number of disintegrations equals the initial activity times its average life, or

the total number of disintegrations during time t

$$= N_0 \lambda_e T_e^a (1-e^{-\lambda_e t}) \qquad (11.11)$$

where N_0 is the initial number of radioactive atoms and T_e^a is the effective average life which is equal to $\frac{T_e^h}{\ln 2}$. Therefore the total dose, in rads, over a time interval t based on days of dwell of the radionuclide in the body is

$$D = 73.8\ C_0 \overline{E}_\beta T_e^h (1-e^{-\lambda_e t}) \text{ rads} \qquad (11.12)$$

where

D = absorbed dose, rad

C_0 = initial concentration, μCi/g

$\overline{E}_\beta$ = average beta energy per disintegration, MeV

T_e^h = effective half-life, day

$e^{-\lambda_e t}$ = fraction remaining at end of time period

Illustration 11.7 Based upon Illustration 11.6, calculate the total dose in rads over a 7-day period. Here Equation (11.12) is appropriate in its use of the effective half-life.

$$\frac{1}{T_e^h} = \frac{1}{T_p^h} + \frac{1}{T_b^h} = \frac{1}{4490} + \frac{1}{12} = \frac{1}{11.97}\ \text{da.}^{-1}$$

$$\text{i.e. } T_e^h = 11.97 \text{ da.}$$

Thus,

$$\begin{aligned} D &= 73.8(15000/43000)(.006)(11.97)(1-e^{-7\ln 2/11.97}) \\ &= 0.616 \text{ rad} \end{aligned}$$

Now consider the dose received due to an internal gamma emitter. In dealing with gamma photons, the radionuclides ingested usually settle in a particular body part called the "critical organ." There the radionuclides act as a distributed source, which must be handled differently from the charged particle dose considered previously.

Again, for a dose calculation, the energy deposited is the quantity of interest. However, since one is dealing with a distributed source, the effects due to both distance and energy absorption from the intensity of gamma photons penetrating the medium must be considered. To account for this effect a quantity called the "geometrical factor," $\bar{g}$, has been derived. Table 11.8 [Reference (11-10)] gives values of $\bar{g}$ for gamma-ray emitters which are uniformly distributed in total body tissue where $\bar{g}$ is based on the body weight (in kg) and the height (in cm). Table 11.9 [Reference (11-20)] gives $\bar{g}$ values for organ shapes which can be represented by cylinders. For geometries spherical in shape, one can use the following to obtain $\bar{g}$

$$\bar{g} = 3\pi R \text{ cm} \qquad (11.13)$$

where

R = radius in cm

To properly obtain the dose rate, one must also consider the distance from the source along with the activity of the source. This is related by the "specific gamma-ray constant," Γ, which is defined as the absorbed dose rate due to a flux density at a distance of one cm from a source activity of 1 mCi (see Table 11.10) [Reference (11-21)].

Relating these two values, one can then calculate the dose rate from an internal gamma emitter as

$$\dot{D} = 0.001\ C\rho\Gamma\bar{g}\ \text{rad/hr} \qquad (11.14)$$

where

C = concentration of the radioisotope in the critical organ, µCi/g
ρ = critical organ density, g/cc
Γ = specific gamma-ray constant
$\bar{g}$ = average geometric factor, cm

Illustration 11.8 30 percent of I-131 is taken up by the thyroid. If the ingested initial activity is 0.4 mCi, what is the absorbed dose rate to the thyroid which has a 2 cm radius and 7.5 cm length?

Based upon Equation (11.14) and Tables 11.9, 11.10, and 11.7, $\bar{g}$ = 23.4, Γ = 2.23 and thyroid mass = 20g. Thus

$$\begin{aligned}\dot{D} &= 0.001\ C\rho\Gamma g\ \text{rad/hr}\\ &= 0.001(400/20)(.3)(2.23)(23.4)\\ &= 0.312\ \text{rad/hr}\end{aligned}$$

TABLE 11.8

Average Geometrical Factor, $\bar{g}$ (cm) for γ-Ray Emitter Uniformly Distributed in Total Body Tissues

Weight kg	Height, cm						
	200	190	180	170	160	150	140
100	138	139	142	145	147	150	154
90	134	136	138	140	143	146	148
80	129	130	131	134	136	139	141
70	123	124	125	126	129	131	135
60	117	118	119	120	122	125	128
50	112	113	114	116	117	119	122
40	102	104	105	106	108	109	110

[Reproduced with permission from J. Shapiro, Radiation Protection, 2nd ed., Harvard University Press, Cambridge, Massachusetts, 1972]

TABLE 11.9

Average Geometrical Factor, $\bar{g}$ (cm) for Cylinders Containing a Uniformly Distributed γ-Ray Emitter

Length of Cylinder, cm	Radius of Cylinder, cm						
	1	2	3	5	10	20	30
2	6.5	11.7	15.7	21.6	25.2	30.5	35.4
5	10.6	18.8	25.6	36.0	48.5	62.6	73.0
10	12.7	23.6	33.0	47.1	70.2	94.0	109
20	14.2	26.7	38.0	56.3	89.6	127	147
30	14.5	27.6	39.7	59.9	98.8	144	172
40	14.8	28.2	40.7	62.4	103	156	187
60	14.8	28.7	41.7	65.5	109	171	206
80	14.8	28.8	42.1	65.8	112	176	214
100	14.8	29.2	42.5	66.2	114	179	218

[From Shapiro, Ref. (11-20)]

TABLE 11.10

Specific Gamma-Ray Constant

Nuclide	Γcm/mCi hr
C-11	5.91
F-18	5.73
Na-22	11.91
Na-24	18.2
Mg-28	16.2
Al-28	16.2
K-42	1.35
Ca-47	6.07
Fe-52	17.2
Fe-59	6.2
Co-57	.93
Co-58	5.52
Co-60	1.29
Cu-64	1.16
As-74	4.33
Rb-86	.49
Tc-99m	.60
I-125	1.66
I-131	2.23
I-132	12.0
Cs-137	3.32
Radium & decay products	8.25

[Reproduced with permission from H.E. Johns and J.R. Cunningham, The Physics of Radiology, 3rd ed., Charles C. Thomas, 1978]

The amount of energy absorbed over a time period is based on the same idea as that covered for the beta dose. The effective half-life is a combination of the physical and biological half-life, and the dose further depends upon the number of disintegrations occurring over the time interval. Therefore, the dose received over a time period t is just

$$D = 0.0346\ C\bar{g}\rho\Gamma T_e^h\ (1-e^{-\lambda_e t})\ \text{rads} \qquad (11.15)$$

where

- C = concentration, μCi/g
- $\bar{g}$ = average geometrical factor, cm
- ρ = critical organ density, g/cc
- Γ = specific gamma-ray constant
- T_e^h = effective half-life, days
- $e^{-\lambda_e t}$ = fraction remaining at end of time period

<u>Illustration 11.9</u> Given the information in Illustration 11.8, what is the dose received after 3 days? Given Equation (11.15), one has

$$T_e^h = (T_p^h)(T_b^h)/(T_p^h + T_b^h) = 60(69)/(60 + 69) = 32 \text{ days},$$

so that

$$D = 0.0346(400/20)(23.4)(.3)(1)(2.23)(32)(1-e^{-3\ \ln 2/32})$$
$$= 21.8 \text{ rads}$$

The material covered in this section was not intended to give an indepth view of dose calculations, but to enlighten the reader on simple methods to estimate the dose received due to an exposure. For a more indepth coverage, see References (11-5) through (11-12).

REFERENCES

(11-1) Cember, H., Introduction to Health Physics, Pergamon Press, 1969.

(11-2) Casarett, A.P., Radiation Biology, Prentice-Hall, Inc., 1968.

(11-3) Hall, E.J., Radiobiology for the Radiologist, 2nd ed., Medical Department, Harper & Row Publishers, 1978.

(11-4) Bacq, Z.M., and Alexander, P., Fundamentals of Radiobiology, 2nd ed., Pergamon Press, New York, 1961.

(11-5) International Atomic Energy Agency, Biological Effects on Ionizing Radiation at the Molecular Level, IAEA, Vienna, 1962.

(11-6) Glasstone, S., and Sesonske, A., Nuclear Reactor Engineering, 3rd ed., D. Von Nostrand Company, New York, 1981.

(11-7) Lamarsh, J.R., Introduction to Nuclear Engineering, Addison-Wesley Publishing Company, Reading, Massachusetts, 1975.

(11-8) Sagan, L., "Radiation and Human Health," EPRI Journal, 1979, pp. 7-14.

(11-9) Luckey, T.D., "Physiological Benefits from Low Levels of Ionizing Radiation," Health Physics, 43, No. 6, 771-789, December, 1982.

(11-10) Glasstone, S. and Sesonske, A., Nuclear Reactor Engineering, D. Von Nostrand Company, Princeton, New Jersey, 1963.

(11-11) "Report of the RBE Committee to the International Commissions on Radiological Protection and on Radiological Units and Measurements," Health Physics, 9, No. 4, 381, April, 1963.

(11-12) Basic Radiation Protection Criteria, National Council on Radiation Protection and Measurements Report 39, 1971.

(11-13) Sagan, L., "Radiation and Human Health," EPRI Journal, September, 1979.

(11-14) BEIR Report, "Biological Effects of Ionizing Radiation," National Academy of Sciences, Washington, D.C., 1980.

(11-15) Permissible Dose for Internal Radiation, International Commission on Radiation Units and Measurements Publication 2, Pergamon Press, New York, 1959.

(11-16) Morgan, K.Z., and Turner, J.E., Principles of Radiation Protection, John Wiley and Son, New York, 1967.

(11-17) Recommendations of the International Commission on Radiological Protection, ICRP Publication 6, Pergammon Press, New York, 1964.

(11-18) Recommendations of the International Commission on Radiological Protection, ICRP Publication 9, Pergamon Press, New York, 1965.

(11-19) Wang, Y., Handbook of Radioactive Nuclides, The Chemical Rubber Company, 1969.

(11-20) Shapiro, J., Radiation Protection, 2nd ed., Harvard University Press, Cambridge, Massachusetts 1972.

(11-21) Johns, H.E., and Cunningham, J.R., The Physics of Radiology, 3rd ed., Charles C. Thomas, 1978.

PROBLEMS FOR CHAPTER 11

1. One person receives a dose of 100mrads of gamma radiation, whereas another person receives 100mrads of proton radiation. Which person has received the higher rem dose?

2. Given the area in which you live, estimate the exposure rate received by a non-radiation worker.

3. A nonshielded point source is being used in an experiment. If the 1 Ci source emits a 1-MeV gamma ray 100 percent of the time, what is the exposure rate 10 feet away?

4. An experiment is being performed which calls for the use of a 2 Ci Au-198 source. What dose rate will the experimenter receive if he is 5 feet from the source?

5. A maintenance crew has been sent to repair a reactor. The fast neutron flux is 2000 neutrons per square centimeter per sec and the thermal neutron flux is 15000 neutrons per square centimeter per sec. What dose rate will the workers receive?

6. Calculate the gamma dose imparted to the tissue after five days to a worker 170 cm tall and with a weight of 70 kg, if 750 μCi of Co-60 has been ingested.

7. Obtain the combined beta and gamma dose rate to the whole body for an 80 kg, 180 cm tall man following the intake of 100 μCi of Na-24.

8. A controlled release of Xe-133 has taken place. If the measured concentration is 0.1 μCi/cc in the surrounding area and the cloud takes two hours to disperse, calculate the following:

   a. the external gamma dose

   b. gamma lung dose (average vol = 3100 cc and the effective radius of the lungs = 10 cm)

   c. the internal beta dose to tissue.

9. About 60 percent of the total iodine in the body is in the thyroid.

   a. Compute the beta dose rate to the thyroid gland weighing 25 grams and having 2 mCi I-131 uniformly distributed throughout the organ.

   b. Compute the gamma dose rate to the thyroid gland. Assume a cylinder 7 cm long and of radius of 2 cm.

10. Based upon Problem 5, how long could the workers remain without exceeding their exposure limits?

12 NEUTRON AND GAMMA RAY DOSIMETRY

This final chapter of the text discusses the very important topic of neutron and gamma ray dosimetry -- the measurement of the neutron and gamma radiation fields. Not only is such information necessary for the accurate interpretation of radiation effects, but also for the protection of personnel working in a radiation environment. This chapter first discusses neutron and gamma ray dosimetry for high level sources (such as would be used in radiation effects studies) and then concludes with low-level detection for personnel monitoring.

12.1 Neutron Dosimetry

For experimental convenience, the neutron energy spectrum is generally divided into three groups:

thermal neutrons* $0 \rightarrow 0.4$ eV
intermediate (or resonance) neutrons 0.4 eV $\rightarrow 0.1$ MeV
fast neutrons 0.1 MeV $\rightarrow \infty$

---

* The fraction of the true room temperature Maxwellian distribution which is above 0.4 eV is only two parts out of a million.

The reason for the 0.4 eV "cutoff" of the thermal neutron region will shortly be presented.

12.1.1 Thermal neutrons. The thermal neutron flux is readily determined by activation of a foil or thin wire of material of relatively large thermal neutron cross section. From Equation (3.52) in Section 3.3 of Chapter 3, the activity of a piece of material of volume V with macroscopic activation cross section Σ_{act} is given as

$$A = \Sigma_{act} \Phi V (1-e^{-\lambda t})e^{-\lambda T} \tag{12.1}$$

where t is the time of exposure, T is the time from the end of the exposure until the activity is determined, and λ is the decay constant of the radioisotope produced by the neutron activation. Thus, measuring the activity of a known "dosimeter foil," such as gold or indium, the neutron flux Φ can be determined. However, in some regions of a reactor, especially near the fuel, there may be large quantities of "epithermal" neutrons, in the form of fission neutrons or neutrons being slowed down. One thus needs to examine how to separate the reaction rate into thermal and non-thermal components. To do this, use is made of the fact that cadmium exhibits a cutoff in its cross section at about 0.4 eV (hence the reason for defining this energy as the upper limit to the thermal neutron region). Most neutrons with energy below 0.4 eV will be absorbed by the cadmium while those with energies greater than 0.4 eV will be transmitted. To illustrate this behavior, Figure 12.1 shows the transmission of a piece of 20-mil-thick cadmium to neutrons striking the cadmium perpendicularly. As seen from this graph there is a fairly sharp cutoff in the transmission below about 0.4 eV. If a gold foil (probe material) is surrounded with cadmium, most of the thermal neutrons will be absorbed in the cadmium rather than the foil and hence the activation of the foil will be due primarily to neutrons above 0.4 eV (epithermal neutrons).

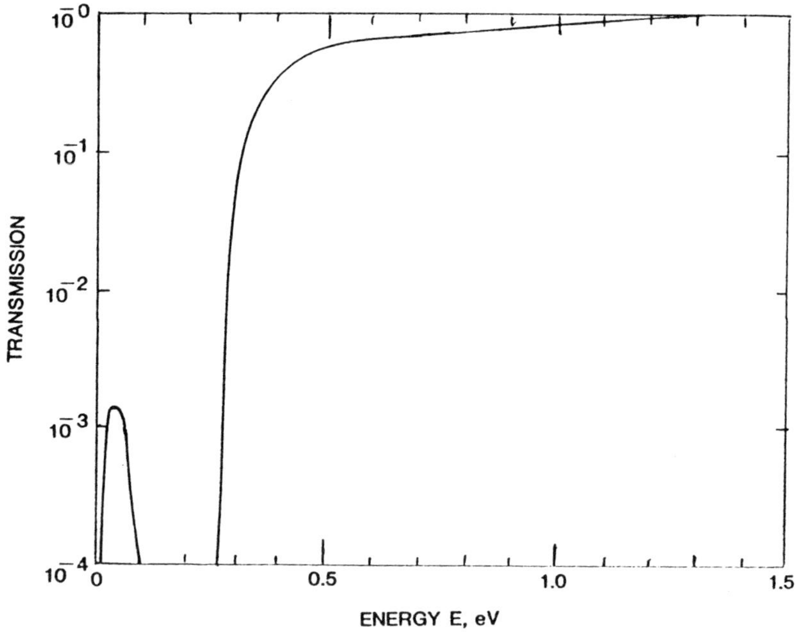

FIGURE 12.1 Transmission of 0.020-inch cadmium for perpendicular neutrons.

By comparison of the activity of the cadmium-covered foil to the activity of a bare foil exposed in the same location, the reaction rate can be separated into that caused by thermal and by epithermal neutrons. Cadmium sleeves are even available for wires of the various dosimeter materials.

Illustration 12.1 A thin gold foil of volume 1 mm^3 is exposed to a pure thermal neutron field at 20°C for one hour. The activity is measured immediately after the exposure to be 1 millicurie. Find the value of the thermal neutron flux.

The absorption cross section for gold-197 is

$$\sigma_a = (98.8)\,\frac{\sqrt{\pi}}{2} \times 10^{-24}\ \text{cm}^2 = 87.9 \times 10^{-24}\ \text{cm}^2$$

The branching ratio for the activation of the gold to gold-198 is 0.99.

Thus

$$\Sigma_{act} = \frac{(19.3)\ (6.02x10^{23})}{(197)}\ (8.79x10^{-23})\ (0.99) = 5.27\ cm^{-1}$$

$$T_{1/2} = 2.7 \text{ days, so that } \lambda = \frac{0.693}{(2.7)(24)} = 0.0107\ hr^{-1}$$

Thus

$$\lambda t = (.0107)(1.0) = 0.0107, \text{ and } (1-e^{-\lambda t}) = 0.011$$

$$A = (10^{-3})(3.7 \times 10^{10}) = 3.7 \times 10^{7} \text{ dis/sec}$$

Thus, from Equation (12.1),

$$\Phi = \frac{(3.7x10^{7})}{(5.27)(10^{-3})(0.011)} = 6.4 \times 10^{12}\ n/(cm^2)(sec)$$

<u>12.1.2 Fast neutrons</u>. Looking at the fast neutron region, to determine the reaction rate one must again have knowledge of $\sigma(v)$ and $n(v)$. Certain materials have a cross section in the fast neutron range approximately a step function, as sketched in Figure 12.2. For these materials the reaction rate can be written

$$R = \int R(v)dv = (\frac{\rho N_A}{A_w}) \int n(v)\ \sigma(v) v\ dv =$$

$$= \frac{\rho N_A \sigma_c}{A_w} \int_{v_T}^{v_{max}} n(v) v\ dv \qquad (12.2)$$

$$= (\frac{\rho N_A \sigma_c}{A_w})(n_T \bar{v}) = (\frac{\rho N_A \sigma_c}{A_w})\ \Phi \quad \text{above threshold velocity } v_T$$

where

$$\sigma_c = \frac{\int_{v \text{ of interest}} n(v)\sigma(v)v\,dv}{\int_{v_T}^{v_{max}} n(v)v\,dv}$$

and

$$\bar{v} = \frac{\int_{v_T}^{v_{max}} n(v)v\,dv}{\int_{v_T}^{v_{max}} n(v)dv}$$

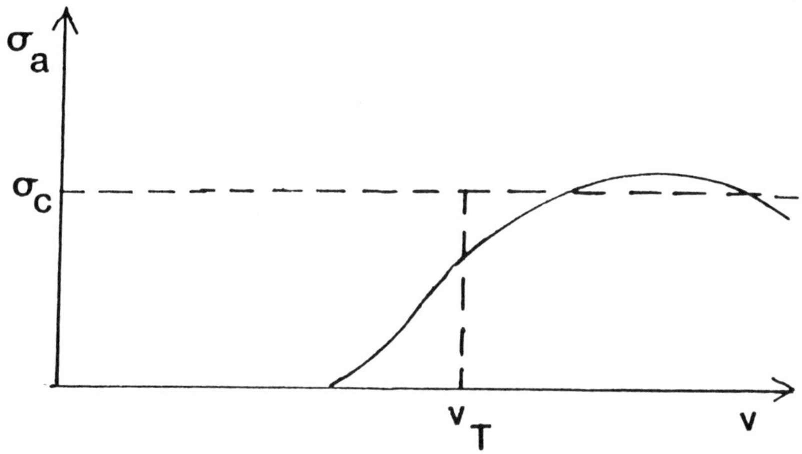

FIGURE 12.2 Sketch of fast neutron cross section of a "threshold" detector material.

In or very near the core, Watt's fission spectrum for fast neutrons can be assumed. Knowing $\sigma(v)$, one can then deduce σ_c, and by activation, determine R, hence Φ. The assumption of a fission spectrum can be checked experimentally by use of "threshold detectors" which have a range of threshold energies. Table 12.1 shows typical threshold reactions.

TABLE 12.1

Typical Threshold Reactions

Reaction	Particle Measured	Threshold Energy, MeV	σ_c, Step Function Cross Section, mb
$^{35}Cl(n,\alpha)^{32}P$	Beta	0.5	--
$^{32}S(n,p)^{32}P$	Beta	2.9	212
$^{27}Al(n,p)^{27}Mg$	Beta	4.6	25
$^{28}Si(n,\alpha)^{25}Al$	1.79 MeV Gamma	5.5	80
$^{24}Mg(n,p)^{24}Na$	Beta	6.2	48
$^{27}Al(n,\alpha)^{24}Na$	Beta	8.1	111

By observing the flux above threshold for each of these reactions, the spectrum can be checked. To obtain the flux above threshold one first assumes Watt's spectrum to calculate σ_c, and then plots the data to check the assumption. Typical data are shown in Figure 12.3 for a thermal research reactor.

12.1.3 Intermediate (resonance) neutrons. Equation (3.2) stated that the resonance region neutron energy spectrum varied as $1/E^{3/2}$. The resonance flux is thus

$$\Phi(E)dE = N(E)E^{1/2}\ dE = \frac{\Phi_0}{E}\ dE \qquad (12.3)$$

i.e., the differential neutron flux is proportional to $\frac{1}{E}$. Thus, to an approximation, one knows the neutron spectrum, but must now know $\sigma(v)$ to complete the determination of the reaction rate.

Unfortunately $\sigma(v)$ is complicated in the resonance range because of the resonances which occur in the cross section. For a material with no resonances, the cross section is still $\frac{1}{v}$ in this range. This behavior is approximately the case for the light elements such as sodium and aluminum. However, most materials have significant resonance peaks superimposed on the $\frac{1}{v}$ behavior. It is here convenient to introduce the concept of the "resonance absorption integral." Note that by substituting Equation (12.3) into the basic equation for reaction rate and changing variable from velocity to energy, one obtains

$$R = \left(\frac{\rho N_A \Phi_0}{A_w}\right) \int_{0.4\ \mathrm{eV}}^{\infty} \sigma(E)\frac{dE}{E} \tag{12.4}$$

where the upper limit has taken as ∞ for convenience, rather than the end of the resonance range (0.1 MeV). Since the behavior is 1/E and no resonances occur above 0.5 MeV, it can readily be seen that

$$\int_{0.4\ \mathrm{eV}}^{0.1\ \mathrm{MeV}} \sigma \frac{dE}{E} >> \int_{0.1\ \mathrm{MeV}}^{\infty} \sigma \frac{dE}{E}$$

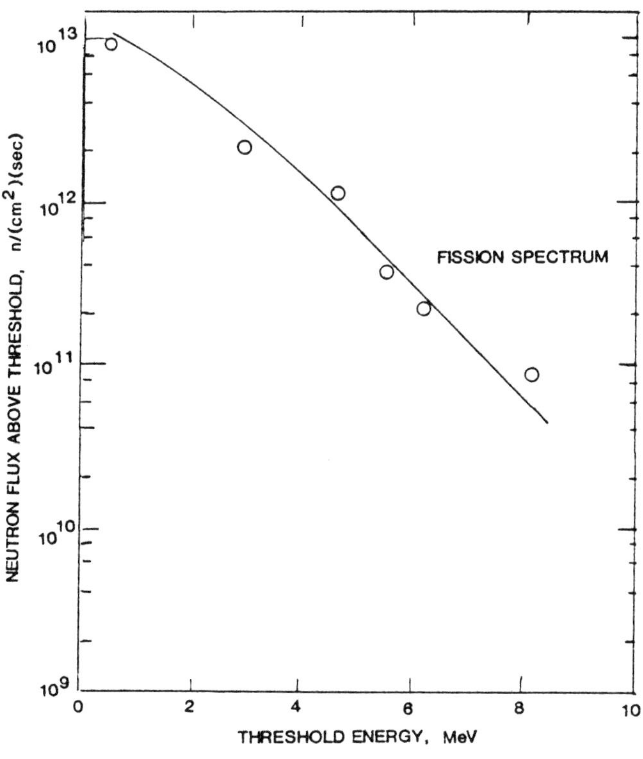

FIGURE 12.3 Fast neutron spectrum at core face of a thermal reactor.

so that the approximation is a good one. When the cross section, σ, is that for absorption, the resonance absorption integral is defined as

$$RI = \int_{0.4\ eV}^{\infty} \sigma_a \frac{dE}{E} \qquad (12.5)$$

This is tabulated for most of the more commonly used reactor materials in Table 12.2. Thus by activation (cadmium-covered) one can determine Φ_0, hence the reaction rate.

The assumption of a $\frac{1}{E}$ flux can be checked experimentally by a method known as "the cadmium difference method." To examine this method, it is convenient to reexamine Equation (12.4), the reaction rate for resonance neutrons. Since the resonance absorption integral has two components (absorption due to the $\frac{1}{v}$ behavior and absorption due to resonances superimposed on it), one may rewrite Equation (12.4) as

$$R = \left(\frac{\rho N_A \Phi_0}{A_w}\right) \left[\int_{0.4\ eV}^{\infty} \sigma_{1/v} \frac{dE}{E} + \int_{0.4\ eV}^{\infty} \sigma_{res} \frac{dE}{E}\right] \qquad (12.6)$$

$$= \left(\frac{\rho N_A \Phi_0}{A_w}\right)(1 + \alpha) \int_{0.4\ eV}^{\infty} \sigma_{1/v} \frac{dE}{E}$$

where

$$\alpha \equiv \frac{\int_{0.4\ eV}^{\infty} \sigma_{res} \frac{dE}{E}}{\int_{0.4\ eV}^{\infty} \sigma_{1/v} \frac{dE}{E}}$$

What will now be shown is that

$$\int_{0.4\ eV}^{\infty} \sigma_{1/v} \frac{dE}{E} = \frac{1}{2}\, \sigma_{2200\ m/sec} \qquad (12.7)$$

TABLE 12.2

Resonance-Absorption Integrals for Common Reactor Structural Materials and Impurities Occurring in These Materials*

Element	Resonance Absorption Integral, barns
Lithium	28
Sodium	0.27
Magnesium	0.9
Aluminum	0.18
Silicon	0.5
Phosphorus	2
Sulfur	0.6
Chlorine	12
Potassium	1.1
Calcium	2
Chromium	1.9
Manganese	10.8
Iron	2.3
Cobalt	48
Nickel	4
Copper	4
Zinc	2
Zirconium	3

* Reference (12-1)

This is done by using $\sigma(v) = \frac{\sigma_0 v_0}{v} = \frac{k}{E^{1/2}}$ and hence

$$\sigma_{2200\ m/sec} = \frac{k}{\sqrt{0.025\ eV}}$$

Substituting the $\frac{1}{v}$ behavior of σ into the left hand side of (12.7),

$$\int_{0.4\ eV}^{\infty} \sigma_{1/v} \frac{dE}{E} = k\int_{0.4\ eV}^{\infty} \frac{dE}{E^{3/2}} = \frac{2k}{\sqrt{0.4\ eV}}$$

$$= \frac{1}{2} \frac{k}{\sqrt{0.025\ eV}} = \frac{1}{2} \sigma_{2200\ m/sec}$$

Hence, substituting into Equation (12.6)

$$R = \left(\frac{\rho N_A \Phi_0}{A_w}\right) (1 + \alpha) \left(\frac{1}{2} \sigma_{2200\ m/sec}\right) \tag{12.8}$$

Thus, if σ is known (from cross-section tables), one has another formula for determining the reaction rate. However, the resonance reaction rate must be measured in the presence of a thermal neutron field. To separate the resonance from the thermal activation one must expose both bare and cadmium-covered foils. The "cadmium ratio" is

$$CR = \frac{\text{reaction rate for bare foil}}{\text{reaction rate for cadmium-covered foil}} = 1 + \frac{\text{thermal reaction rate}}{\text{resonance reaction rate}} \tag{12.9}$$

Rearranging and making substitutions

$$CR-1 = \frac{\text{thermal R.R.}}{\text{resonance R.R.}} = \frac{\left(\frac{\rho\ N_A}{A_w}\right)(n_T v_{2200})\sigma_{2200}}{\Phi_0(1+\alpha)\left(\frac{\rho\ N_A}{A_w}\right) \frac{1}{2}\sigma_{2200}}$$

$$= \frac{2(n_T v_{2200})}{\Phi_0(1+\alpha)} = \frac{2\ \Phi_{2200}}{\Phi_0(1+\alpha)}$$

or hence

$$\Phi_0 = \frac{2\ \Phi_{2200}}{(CR-1)(1+\alpha)} \tag{12.10}$$

Thus, in a fairly elaborate way one has once again arrived at how to obtain Φ_0. From Equation (12.4) one can determine Φ_0 from a single measurement with a cadmium-covered foil, knowing the resonance absorption integral. However, the significant fact about Equation (12.10) is the presence of α, the ratio of the reactions due to the resonances alone to those due to the $\frac{1}{v}$ part of the cross section behavior. By use of special materials, "resonance detectors," by use of α one can check the assumption of a $\frac{1}{E}$ flux spectrum in the resonance range. Resonance detectors are materials for which α is due principally from one resonance peak at energy E_R. By performing measurements with various materials which have different E_Rs, one can solve Equation (12.10) for Φ_0 for each E_R. If the Φ_0s thus obtained are the same for all E_R, then the spectrum can be said to be $\frac{1}{E}$ over the energy range investigated. If the Φ_0s are not approximately the same, little can be said about the spectrum.

A study of the differential neutron flux $\Phi(E)$ was made in the same location as used for the fast neutron flux data shown in Figure 12.3. The detectors and data for a particular experiment are shown in Figure 12.3. It is seen that the Φ_0s are approximately the same over this energy range. These data are plotted in Figure 12.4 to demonstrate better the $\frac{1}{E}$ behavior. The solid line of the graph is the true $\frac{1}{E}$ spectrum.

TABLE 12.3

Differential Neutron Flux in the Energy Region of 260 eV to 3000 eV

Reaction	E_R, Resonance Energy, eV	α	CR	Φ_0
$^{55}Mn(n,\gamma)^{56}Mn$	260	1.33	14.4	3.53
$^{63}Cu(n,\gamma)^{64}Cu$	570	0.933	21.5	2.77
$^{37}Cl(n,\gamma)^{38}Cl$	1800	0.079	30.4	3.47
$^{51}V(n,\gamma)^{52}V$	3000	0.134	37.6	2.65

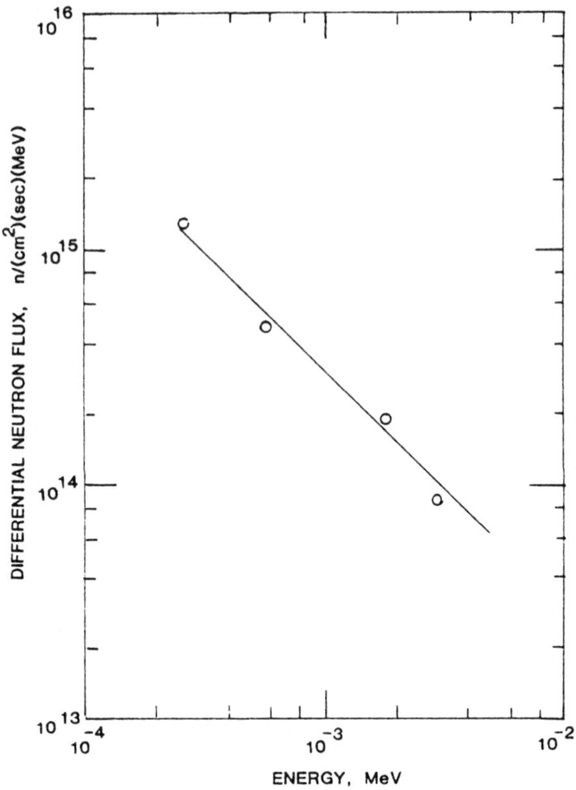

FIGURE 12.4 Differential-resonance neutron flux at the core face of a thermal research reactor. (Battelle Memorial Institute)

12.2 Gamma Ray Dosimetry

Of particular interest to a study of radiation effects is "high-level" gamma radiation measurement. During the past several decades considerable progress has been made in the development of instruments and techniques for measuring high gamma dose rates. One of the earlier instruments, the ionization chamber, has been greatly improved and newer devices, such as thermoluminescent detectors, have been commercialized.

12.2.1 Ion chambers. The ion chamber detects radiation as a result of the ionization produced by the radiation as it passes through matter. The chamber is a gas-filled tube containing a well-insulated central wire (collecting electrode) which is usually maintained at a positive electrical DC potential with respect to the cylinder wall (cathode). When radiation passes through the tube, the gas molecules are ionized. The free electrons are rapidly accelerated toward the collecting electrodes as a result of the attraction of unlike charges, while the more massive positive ions move at a slower rate toward the cathode. When the ions reach the electrodes, they are neutralized by charges supplied by the voltage source, producing a current through the external circuitry, voltage generator, collecting electrode, gas and cathode. The current, or voltage pulses produced by the current in the external circuitry, is a measure of the radiation which induced it.

For a given radiation intensity, the current produced is a sensitive function of applied voltage. At very low voltages, the electrons and ions move relatively slowly, and have a chance to recombine before reaching the electrodes. This recombination neutralizes the ions before they are collected. This results in a reduced current output. As the applied voltage increases, the recombination decreases, resulting in an increased current output. Eventually, all of the electrons and ions produced by the radiation are collected, and the current remains constant as the voltage is increased. This region is known as the ionization region, and when gas detectors are operated in this region, they are known as ionization chambers. The current at which this phenomenon occurs is known as the "saturation current." In the ionization region, the number of ions collected have a one-to-one correspondence with the number of ions produced by the radiation (primary ions). This occurs over a relatively wide voltage band.

If the voltage is increased beyond the ionization region, however, the ionized electrons have sufficient energy to ionize other molecules with which they collide, producing secondary electrons and ions. This phenomenon is known as multiplication, and when gas detectors are operated in this voltage region, they are known as proportional counters. The current then becomes much more sensitive to the applied voltage.

When operation is maintained in the ionization region the chamber is stable over a relatively large voltage band, and produces currents which are directly proportional to the radiation intensity incident upon them. The ion chamber can be constructed in many different sizes and shapes, and can use a variety of filler gases. Their sensitivity to radiation can be varied by varying the gas pressure within the chamber, and they can provide instantaneous readings of dose or exposure rate. These properties make the ion chamber an excellent choice for dosimetry measurements. When vented to the atmosphere, the ion chamber becomes an exposure meter which reads in Roentgens when the incident radiation is in the form of X- or gamma rays, since the Roentgen is defined in terms of the ionization produced in air by electromagnetic radiation.

For the measurement of MegaRoentgens per hour exposure rates, which is common in certain applications, care must be taken to limit the total exposure to values below 10 MegaRoentgens. Otherwise, the plastic insulation commonly used on present-day designs will break down and spurious readings will result. To prevent pulse pileup at such high radiation intensities, chambers with very small active volumes are used. Typically, volumes less than a cubic centimeter are utilized. As the rate of production of ions increases it becomes increasingly difficult for the ion chamber to collect all of the ions produced, and the current saturates as radiation intensity increases.

Recent experiments at the Cobalt-60 Laboratory of the University of Cincinnati showed exposure rates as high as 60,000 Roentgens per hour did not saturate a 0.33 cc atmospheric chamber. Excellent linearity was demonstrated up to those exposure levels. Figure 12.5 shows a typical axial dose rate curve determined by an ion chamber. This type of measurement illustrates the utility of the ion chamber in obtaining instantaneous dose rate data.

In summary, the ion chamber can be used for a wide variety of applications as a result of its inherent flexibility, but care must be taken not to overexpose it in high dose applications.

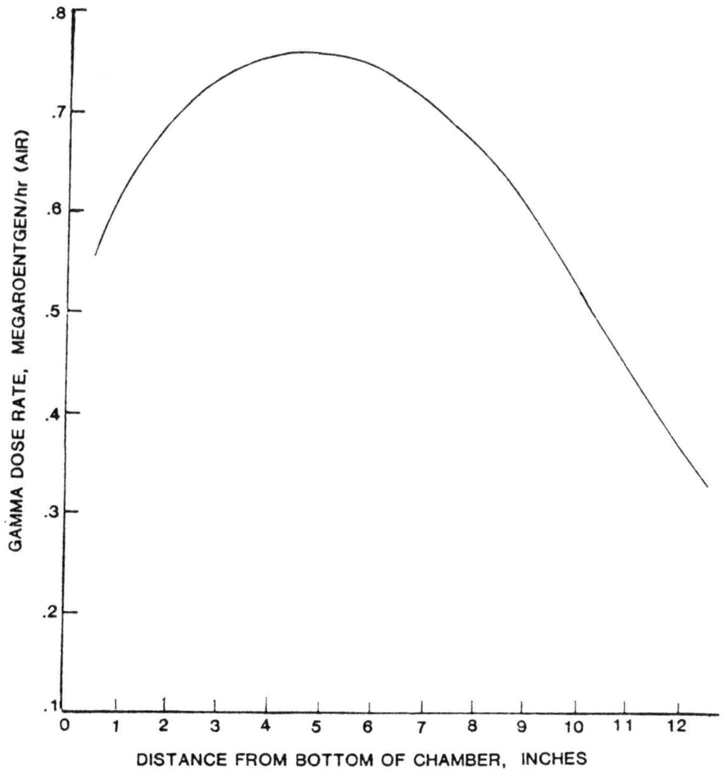

FIGURE 12.5 Axial measurement of gamma dose rate in COBALT-60 irradiation facility (by ionization chamber).

12.2.2 TLD dosimeters. Thermoluminescent dosimeters (TLDs) are solid crystals with impurity atoms. Radiation raises the energy of the electrons in the crystal so that they are freed from their atoms. The free electrons are eventually caught in the imperfections, called traps, at a higher energy state than they were originally. When heated, they are freed from their traps and return to their original energy (the valence band) emitting light. The light can be converted to electrical currents by a photomultiplier tube, and recorded. The number of trapped electrons and, therefore, the light emitted is proportional to the amount of radiation incident on the dosimeter. Thus the TLD dosimeter is an integrating device which records total dose.

At present the most popular crystalline material for TLD use is lithium fluoride, with the addition of manganese and titanium impurity atoms (activators). LiF shows good linearity of light output with dose, reasonably constant energy response above 0.1 MeV for gamma rays, adequate beta ray response, resistance to fading with time, reusability after annealing, and good reproducibility. Another useful material is calcium fluoride with a manganese activator.

The energies at which electrons are trapped vary as a result of variations in impurity states. Thus, some electrons will be released at lower temperatures (energies) than others. A plot of light output versus temperature is known as a glow curve. It shows peaks at those temperatures corresponding to trap energies. In the normal mode of operation, the integrated light output under the glow curve, which is directly proportional to dose, is read and recorded.

The common TLDs exhibit saturation of the integrated light output at doses above about 100,000 rads. In essence, all of the trap sites become occupied at large doses, so additional doses do not produce an increase in trapped electrons, or light output when the dosimeter is heated. Thus the common TLDs and associated techniques cannot be used for megarad dose measurements. Various other techniques have been devised to overcome the saturation of the common TLD materials. Some optical grade LiF powders produce thermoluminescence when heated at very high temperatures (450 degrees Celsius) after being exposed to megarad doses and above. Their light output can be calibrated against known doses. Another technique is to calibrate the ratio of two or more high temperature peaks in ordinary LiF crystals. Additional peaks show up at high exposures and their ratios can be calibrated against dose. While these techniques can be made to work, they are relatively difficult to apply.

12.2.3 Optical methods. In certain materials radiation produces opaqueness at high dose levels and at preferential wavelengths. Optical grade LiF crystals have been developed which yield accurate doses at very high dose levels, beginning in the megarad range. They are relatively easy to use, since their optical transmittance or absorbance can be accurately measured with any good laboratory photospectrometer. Once their transmittance is calibrated against known doses at selected wavelengths, they will provide repeatable dose values. They can be annealed, and therefore reused. They can be constructed in various shapes and sizes to fit most applications. The optical measurements are considerably easier than the temperature measurements associated with TLD dosimetry, so this technique appears to be the most promising for routine high dose dosimetry.

Another optical method, which is NBS traceable, is the "dye film" technique. Dye films are radiochromic chemicals in nylon films. Upon irradiation, the films become permanently colored, with strong absorption bands occurring at definite wavelengths. The optical density at the wavelength increases with radiation and, with proper calibration, provides a measure of the radiation.

12.2.4 Gamma heating dosimetry. A much-neglected method of dosimetry for high-level gamma radiation is a simple gamma heating method based upon the transient temperature behavior of a small sample of material of known specific heat. If the measurement is performed properly, accuracies in dose rate determination can be well within ± 5 percent [References (12-2) through (12-5)].

Consider a small body whose temperature is sufficiently uniform to be considered independent of the position within the body. Assume that heat is lost from the body only by heat transfer from the surface and that the rate of loss of heat is proportional to the temperature difference between the body and its surroundings. The net change in heat content of this body, upon heating, can be expressed in the form

$$V\rho c_p \frac{dT}{dt} = gV - hST(t) \qquad (12.11)$$

where

V = volume of the body
ρ = density
c_p = specific heat
T = temperature
t = time
g = heat generation rate per unit volume due to gamma radiation
h = heat transfer coefficient
S = surface area effective in heat transfer

Note that the temperature of the surroundings has been

arbitrarily chosen to be zero. The left hand side of Equation (12.11) is the time rate of change of heat content of the body. The term gV represents the heat generation rate in the body, and hST is the rate of external heat loss. Let $\lambda = \frac{hS}{\rho V c_p}$, then Equation (12.11) becomes

$$\frac{dT}{dt} = \frac{g}{\rho c_p} - \lambda T = \lambda \left(\frac{g}{\rho c_p \lambda} - T\right) \tag{12.12}$$

Again, let $\mu \equiv \rho c_p \lambda$, then

$$\frac{dT(\theta)}{dt} = \lambda \left(\frac{g}{\mu} - T\right) \tag{12.13}$$

Upon setting the initial temperature of the body as zero, the temperature at time t is given as

$$T(t) = \frac{g}{\mu}\left(1-e^{-\lambda t}\right) \tag{12.14}$$

This is the gamma heating equation for the body. Note that after a long time, the temperature will reach equilibrium and has a value of $\frac{g}{\mu}$.

Now consider the same body at initial temperature T_0. With no heat generation, the temperature has the form

$$\frac{dT(t)}{dt} = -\lambda T(t) \tag{12.15}$$

Hence the temperature at time t is given by

$$T(t) = T_0 e^{-\lambda t} \tag{12.16}$$

This is the cooling equation. Thus, if the body is heated to a temperature T_0 and then the heat source is removed and the body is allowed to cool, the value of λ can be found from a semilog plot of the cooling curve.

With the λ found from the cooling curve, a line of

$T(t)$ versus $1-e^{-\lambda t}$ can be plotted to yield a slope of $\frac{g}{\mu}$. Since $\frac{g}{\mu} = \frac{g}{\rho c_p \lambda}$, the gamma heating rate per unit mass, g/ρ, can be readily found once the value of c_p is given. Note that $\frac{g}{\rho}$ has the unit of joule/(sec)(gm) which can be directly converted to a rad/hr dose rate.

In this analysis, three principal conditions have to be met:

1. the rate of loss of heat must be proportional to the temperature difference between the body and its surroundings,

2. the temperature of the body must be uniform (or at least the deviation from uniformity must be within the desired accuracy of the measurement), and

3. the temperature of the surroundings must meet the same requirements on uniformity as the body.

Condition (1) is met for the heat transfer by conduction if the temperature changes are small enough to avoid changes in thermal conductivity. It is also met approximately for heat transfer by radiation since $q = 4KT^3 \Delta T$ if the temperature difference, ΔT, is small compared to T. However, heat transfer by free convection does not meet the condition. Various correlations indicate that the convective heat-transfer coefficient is approximately proportional to $(\Delta T)^{1/4}$. Consequently, it is necessary to design transient-temperature devices such that heat transfer by convection is negligible compared to conduction and radiation. This is done by pumping a mild vacuum around the specimen.

By attaching a thermocouple to a small sample of material of known specific heat, measurement of the heating and cooling curves in a small evacuated chamber will determine the heat generation rate, hence, the gamma dose rate. A sketch of such a device is shown in Figure 12.6. However, the cooling curve only needs to be determined once (initially). This provides the value of λ, which is a constant of the device. One can then analyze the slope of the heating curve in a gamma radiation field, which is given from Equation (12.14) as

$$\frac{dT}{dt} = \frac{\lambda g}{\mu} e^{-\lambda t} = \frac{g}{\rho c_p} e^{-\lambda t} \tag{12.17}$$

(In fact, if one analyzes the initial slope

$$\left(\frac{dT}{dt}\right)_{t=0} = \frac{g}{\rho c_p} \tag{12.18}$$

the analysis is independent of λ). Since $(\frac{g}{\rho})$ is the rad dose rate, one must only have an NBS traceable determination of the specific heat to obtain the dose rate. Thus, one places the device of Figure 12.6 into a gamma field, analyzes the rate of temperature rise, and prints out or displays the gamma dose rate (after appropriate conversion of units).

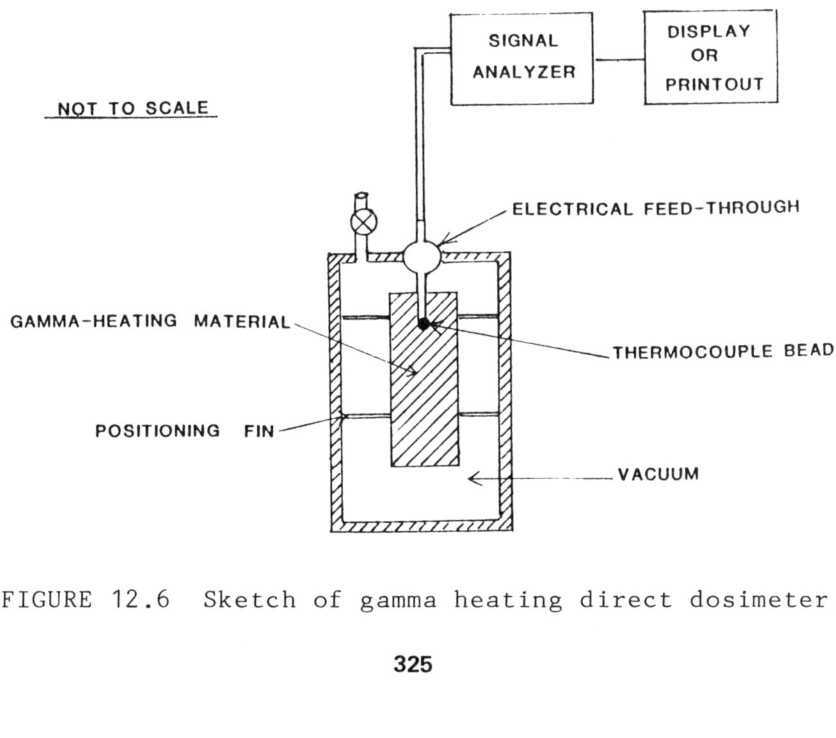

FIGURE 12.6 Sketch of gamma heating direct dosimeter.

12.3 Health Physics Dosimetry

In addition to dosimetry measurements required to properly interpret radiation effects on materials and components, an important area of dosimetry is in health physics -- the monitoring of personnel working in radiation environments. This section of Chapter 12 briefly discusses several types of monitors used for this purpose.

12.3.1 Film badges. The film badge was one of the earliest monitors attached to personnel to determine their exposure and is still widely used in the industry. The principle of the film badge is that the film darkens under radiation exposure, the degree of darkening being related to the total radiation dose received. By photometrically determining the darkness of the film, the exposure can be determined. With appropriate filters, the radiation exposure can be separated into that from gamma (penetrating), beta (non-penetrating), and neutrons. Film badges can be purchased in various types: whole body (usually clipped to the waist or shirt pocket); ring badges, for use in situations where the hands might receive most of the exposure; wrist badges and eye or collar badges. Typically, the film in the badges is measured on a monthly or bimonthly basis, and the results are recorded on an employee's permanent exposure record. Figure 12.7 shows a typical film badge holder with identifying labelling [Reference (12-6)]. The sensitivity of the film badge is of the order of 10 mrem exposure.

12.3.2 TLD personnel dosimeters. The principle of the thermoluminescent detector (TLD) used for high-level gamma dosimetry as discussed in Section 12.2, can be applied to personnel dosimetry. In fact, TLD personnel dosimeters are rapidly gaining favor over film badges. An example of the properties of the TLD gamma dosimeter is given by the Landauer TLD ring dosimeter [Reference

(12-6)]. This dosimeter is made of thermoluminescent (TL) grade lithium fluoride. It has an exposure range of 10 mR to 10^5 R. The standard error for gamma radiation is less than 25% or 10 mR, whichever is greater from 10 mR to 100 mR, and is ± 10% above 100 mR. The TL dosimeter is dose rate independent up to at least 2 x 10^8 R/sec, and its response is energy independent for X- and γ- radiations down to 100 KeV, with fairly small energy dependence for lower energies down to 10 KeV.

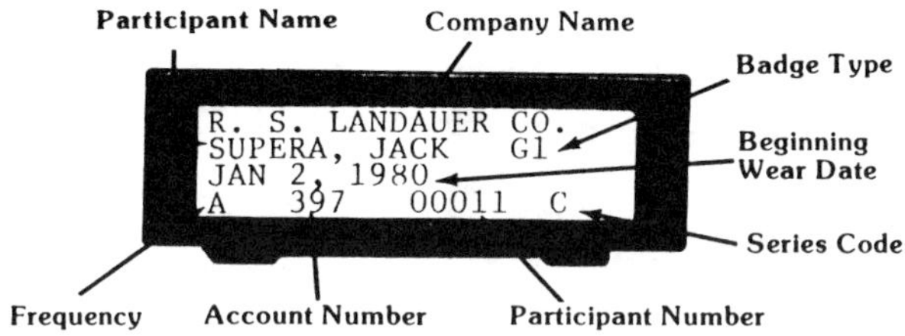

FIGURE 12.7 Typical film badge holder with identifying labelling.

[Reproduced with permission from Landauer Film Badge Brochure, R.S. Landauer, Jr. and Co., Division of Tech/Ops, Inc., Glenwood, IL, 1983]

<u>12.3.3 Pocket dosimeters</u>. Direct-reading pocket dosimeters are in common use for immediate indication of radiation exposure. These are essentially miniature air ionization chambers connected to a quartz fiber electrometer. Figure 12.8 shows both a picture and a schematic of a pocket gamma dosimeter [Reference (12-7)]. These are commercially available in a variety of ranges. Probably the most common is the 0-200 mR range model, but they can be obtained with ranges as high as 0-600 R. Typical accuracy is ± 10% at midscale. In early versions of the pocket ion chamber, drift of the indicator was a problem. However, modern versions meeting ANS specifications have drift rates of less than 2% per day at room temperatures [Reference (12-7)]. By modifying the chamber with boron lining, a direct-reading thermal neutron dosimeter results.

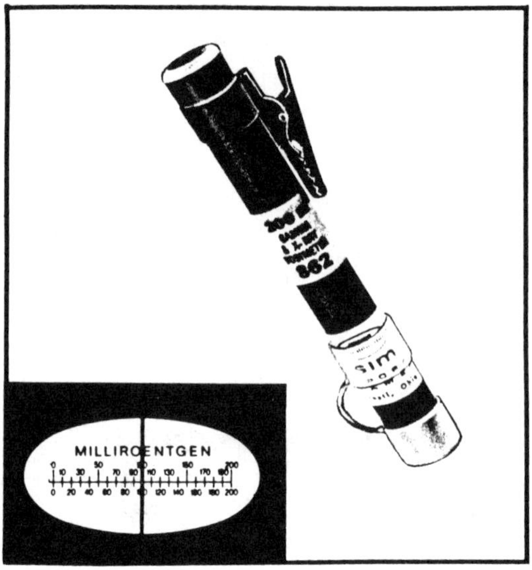

FIGURE 12.8(a) Photograph of pocket gamma detector.
[Reproduced with permission from Dosimeter Corporation Direct-Reading Dosimeter Corporation, Cincinnati, Ohio, 1983]

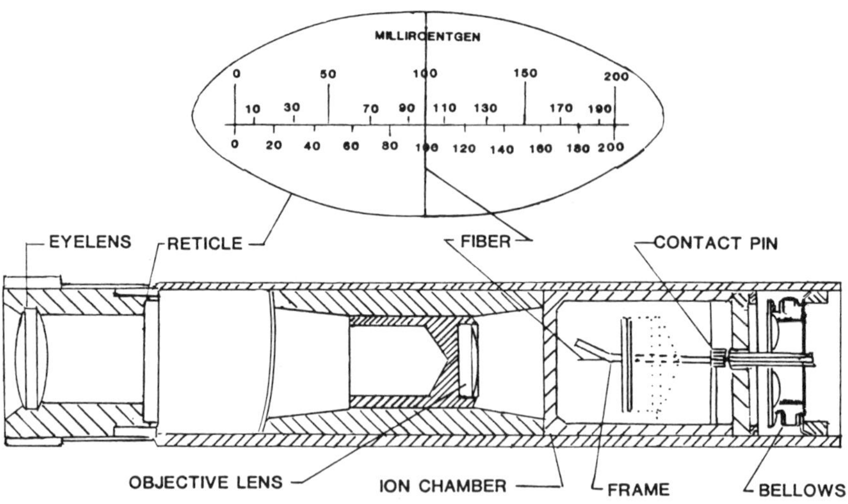

FIGURE 12.8(b) Schematic of pocket gamma detector.
[From Dosimeter Corporation, Ref. (12-7)]

Dosimeter Corporation has conceived a particularly intriguing computerized arrangement for their direct-reading pocket dosimeters [Reference (12-8)]. A system schematic is shown in Figure 12.9. One places the pocket dosimeter in a holder, a camera views the readout, which then gets converted to a signal for computer display, printout and/or storage. With this system, one has a personnel safety system (or entry control system) which requires the worker to obtain a computer reading before entering or exiting the place of employment. The data can then automatically be monitored and stored as a part of the worker's exposure record.

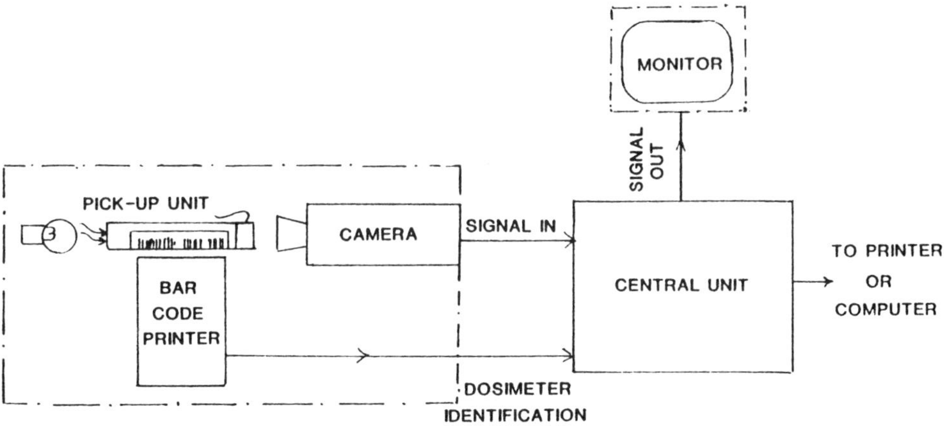

FIGURE 12.9 Schematic for computerized direct-reading dosimeter readout system.

[Reproduced with permission from Dosimeter Corporation Brochure, "Computer Interfacing Automatic Dosimeter-Reader (CID)," Dosimeter Corporation, Cincinnati, Ohio, 1983]

REFERENCES

(12-1) Nucleonics, Data Sheet No. 23, March, 1958.

(12-2) Anno, J.N., "Fission Plate Power Measurements by a Transient-Temperature Method," Nuclear Science and Engineering, 16, 357-362, 1963.

(12-3) Epel, L.G., "Transient Temperature in Infinite Plates, Infinite Cylinders, and Spheres Following a Simultaneous Step Change in Internal Heat Generation Rate, Coolant Temperature and Heat Transfer Coefficient," ORNL-2597, November 14, 1958, Oak Ridge National Laboratory, Oak Ridge, Tennessee.

(12-4) Epel, L.G., "Transient Temperatures in Infinite Plates, Infinite Cylinders and Spheres During Convective Cooling from Initially Parabolic Temperature Profiles," ORNL-2511, May 5, 1958, Oak Ridge National Laboratory, Oak Ridge, Tennessee.

(12-5) Binford, F.T., Bettis, E.S., and Howe, J.T., "Gamma Heating Measurements in the Bulk Shielding Reactor," Report CF-56-3-72, 1956, Oak Ridge National Laboratory, Oak Ridge, Tennessee.

(12-6) Landauer TLD Ring Brochure, R.S. Landauer, Jr. and Company, Division of Tech/Ops, Inc., Glenwood, Illinois, 1983.

(12-7) Dosimeter Corporation Direct-Reading Dosimeter Brochure, Dosimeter Corporation, Cincinnati, Ohio, 1983.

(12-8) Dosimeter Corporation Brochure, "Computer Interfacing Automatic Dosimeter-Reader (CID)," Dosimeter Corporation, Cincinnati, Ohio, 1983.

PROBLEMS FOR CHAPTER 12

1. The average epithermal (thermal to 0.1 MeV) neutron flux in the Zimmer core is 8.45×10^{13} n/(cm^2)(sec).

   a. Assuming the differential epithermal neutron flux $\Phi(E)$ varies with energy as

   $$\Phi(E)dE = \frac{C}{E}\,dE$$

   where E is the neutron energy and C is a constant, determine C.

   b. What is the average energy $\overline{E}$ of these epithermal neutrons?

   Hint: $$\overline{E} = \frac{\int N(E)E\,dE}{\int N(E)\,dE}$$

2. A cylinder of lead (Pb) with dimensions of 0.375 in. diameter by 0.500 in. length is exposed in vacuum to a gamma dose rate of 2×10^7 R/hr (air equivalent). What is the initial rate-of-rise of the temperature?

APPENDIX A: NUCLEAR DATA FOR VARIOUS ELEMENTS AND ISOTOPES

Atomic No. Z	Element or Isotope	Abundance %	Atomic Mass Weight u	Density gm/cm^3	Half-Life	Neutron Cross Sections (barns): Absorption at 0.025eV	Scattering: Thermal	Scattering: Epithermal
1	H		1.00797			0.332	38	20.4
	^{1}H	~100	1.007825			0.332	38	20.4
	^{2}H(D)	0.0151	2.01410			0.00046	7	3.4
	^{3}H(T)	--	3.01605		12.6y			
2	He		4.0026			0.007	0.8	0.83
	^{3}He	0.00013	3.01603			5500	0.8	
	^{4}He	~100	4.00260			0	0.8	
3	Li		6.939	0.53		70	1.4	0.9
	^{6}Li	7.52	6.01513			945(n,α)		
	^{7}Li	92.48	7.01601			0.033		
	^{8}Li	--			0.845s			
4	^{8}Be	--	8.00531		~3×10^{-16} s			
	^{9}Be	100	9.01219	1.82		0.010	7	6.11
5	B		10.811	2.54		755	4	3.7
	^{10}B	19.8	10.01294			3813(n,α)		
	^{11}B	80.2	11.00931			$<50\times10^{-3}$		
	^{12}B	--			0.019s			
6	C		12.01115	2.22		0.0034	4.8	4.66
	^{12}C	98.89	12.00000					
	^{13}C	1.11	13.00335			0.0005		
	^{14}C	--			5570y			
7	N		14.0067			1.88	10	9.9
	^{13}N	--			10m			
	^{14}N	99.63	14.00307			1.75		
	^{15}N	0.37	15.00011					
	^{16}N	--			7.4s			
8	O		15.9994			0.00019	4.2	3.75
	^{16}O	99.759	15.99491					
	^{17}O	0.037	16.99914					
	^{18}O	0.204	17.99916					
	^{19}O	--			29s			
9	^{19}F	100	18.99840			0.010	3.9	3.6
11	^{23}Na	100	22.98977	0.971		0.53	4.0	3.1
	^{24}Na	--		15.05h				
12	Mg		24.312	1.74		0.063	3.6	3.4
	^{24}Mg	78.60	23.98504			0.034		
	^{25}Mg	10.11	24.98584			0.280		
	^{26}Mg	11.29	25.98259			0.060		
	^{27}Mg	--			9.45m			
13	^{27}Al	100	26.9815	2.70		0.230	1.4	1.4
	^{28}Al	--			2.3m			
14	Si		28.086	2.4		0.16	1.7	2.2
	^{28}Si	92.27	27.97693			0.080		
	^{29}Si	4.68	28.97649			0.280		
	^{30}Si	3.05	29.97376			0.40		
	^{31}Si	--			2.62y			
15	^{31}P	100	30.97376	2.34		0.20	5	3.4
17	^{37}Cl	24.47	36.96590			56		
18	^{37}Ar	--	36.96678		35.1d			
19	K		39.102	0.87		2.07	2.5	2.1
22	Ti		47.90	4.5		5.8	4	4.2
23	V		50.942	6.0		4.5	5.0	
24	Cr		51.996	6.92		3.1	3.0	3.9
	^{53}Cr	9.55	52.9407			1.82		
25	^{55}Mn	100	54.9381	7.2		13.2	2.3	1.9
	^{56}Mn	--			2.58h			
26	Fe		55.847	7.87		2.62	11	11.4
	^{54}Fe	5.84	53.9396			2.3		
	^{55}Fe	--			2.6y			
	^{56}Fe	91.68	55.9349			2.7		
	^{57}Fe	2.17	56.9354			2.5		
	^{58}Fe	0.31	57.9333			1.2		
	^{59}Fe	--			45d			
	^{58}Co	--			71.3d	1.9×10^{3}		
27	^{59}Co	100	58.9332	8.71		18&19	7	5.8
	^{60}Co	--			10.5m&5.24y			
28	Ni		58.71	8.9		4.8	17.5	17.4
29	Cu		63.54	8.96		3.77	7.2	7.7
	^{63}Cu	69.1	62.9298			4.6		
	^{64}Cu	--			12.9h			
	^{65}Cu	30.9	64.9278			2.2		
	^{66}Cu	--			5.1m			
34	^{79}Se	--			3.89m&6.5×10^{4} y			
35	^{79}Br	50.69	78.91833			2.48&8.4	6	6

36	^{79}Kr	--			50s&34.9h			
37	^{87}Rb	27.85			4.7×10^{10} y	0.12		
40	Zr		91.22	6.44		0.180	8	6.2
41	^{93}Nb	100	92.90638	8.57		1.15	5	6.5
	^{94}Nb	--			6.6m			
42	Mo		95.911	10.22		2.7	7	6
47	Ag		107.870	10.50		63	6	6.4
	^{107}Ag	51.35	106.9051			31		
	^{108}Ag	--			2.3m			
	^{109}Ag	48.65	108.9047			87		
	^{110}Ag	--			24s			
48	Cd		112.40	8.64		2450	7	
49	In		114.82	7.28		196	2.2	
	^{113}In	4.2	112.9043					
	^{114}In	--			72s&49d			
	^{115}In	95.77	114.9039		6×10^{14} y	50and150		
	^{116}In	--			14s and 54m			
54	Xe		131.30			35	4.3	
	^{135}Xe	--			92h	2.72×10^{6}		
	^{136}Xe	8.87	135.9072			0.15		
62	Sm		150.35	7.7		5600		
	^{149}Sm	13.84	148.9169			40,800		
64	Gd		157.25	7.94		46,000		
	^{155}Gd	14.73	154.9226			61,000		
	^{157}Gd	15.68	156.9239			240,000		
73	^{181}Ta	~100	180.9480	16.6		21	5	6
	^{182}Ta	--			115d	8200		
79	^{197}Au	100	196.9666	19.3		98.8	9.3	
	^{198}Au	--			2.7d	26,000		
82	Pb		207.19	11.35		0.170	11	11.3
	^{204}Pb	1.5	203.9731		1.4×10^{17} y	0.8		
	^{205}Pb	--			3×10^{7} y			
	^{206}Pb	23.6	205.9745			0.025		
	^{207}Pb	22.6	206.9759			0.70		
	^{208}Pb	52.3	207.9766			0.03		
	^{209}Pb	--	208.9810		3.3h			
83	^{209}Bi	100	208.9804	9.8		0.019&0.015	9	9.28
	^{210}Bi	--			2.6×10^{6} y&50d			
	^{211}Bi	--			2.15m			
84	^{210}Po	--	209.9829		138d			
	^{213}Po	--	212.9928		4μs			
88	^{226}Ra	--	226.0254		1620y	20		
90	^{230}Th	--	230.0331		7.6×10^{4}y			
	^{232}Th	100	232.0382	11.5	1.45×10^{10} y	7.56	12.5	12.5
	^{233}Th	--			22.1m			
91	^{233}Pa	--			27.4d	43		
92	U		238.03	19.0		7.68	8.3	
	^{233}U		233.0395		1.62×10^{5} y	530(n,f) 46(n,γ)		
	^{234}U	0.0057	234.0409		2.5×10^{5} y	105		
	^{235}U	0.714	235.0439		7.1×10^{8} y	582(n,f) 112(n,γ)	10	
	^{236}U	--	236.0457		2.39×10^{7} y	7		
	^{237}U	--			6.75d			
	^{238}U	99.28	238.0508		4.51×10^{9} y	2.71	8.3	
	^{239}U	--			23.5m	14(n,f)		
93	^{239}Np	--			2.35d			
	^{238}Pu	--	238.04958		87.8y			
94	^{239}Pu	--	239.0522	19.6	24,360y	746(n,f) 280(n,γ)	9.6	
	^{240}Pu	--	240.0540		6760y	<0.1(n,f) 295(n,γ)		
	^{241}Pu	--			15y	1025(n,f) 375(n,γ)		
	^{242}Pu	--	242.0587		3.79×10^{5} y	<0.2(n,f) 30(n,γ)		
95	^{241}Am	--	241.0567		433y	710(n,γ) 3(n,f)		
96	^{242}Cm	--	242.0588		163d	25(n,γ) <5(n,f)		
	^{244}Cm	--	244.0628		18.1y	~13(n,γ) 1(n,f)		
98	^{252}Cf	--	252.08		2.65y(α) 85.5y(s.fis.)	20(n,γ)		

[From Arthur R. Foster and Robert L. Wright, Jr., Basic Nuclear Engineering, 3rd ed., Copyright (c)1977 by Allyn & Bacon, Inc., reprinted with permission]

APPENDIX B: DERIVATION OF THE FRACTION OF ALL BETAS ESCAPING FROM THE SURFACE

Assume linear paths of betas in the material and isotropic emission, i.e., $N_0/4\pi$ per steradian. As seen in Figure B.1, the fraction of betas emitted from depth x is

$$F = \frac{\int_0^{2\pi} \int_0^{\cos^{-1}\frac{x}{R}} \frac{N_0}{4\pi} \sin\theta d\theta d\Phi}{\int_0^{2\pi} \int_0^{\pi} \frac{N_0}{4\pi} \sin\theta d\theta d\Phi}$$

$$= \frac{\int_0^{\cos^{-1}\frac{x}{R}} \sin\theta d\theta}{\int_0^{\pi} \sin\theta d\theta}$$

$$= \frac{- \cos\theta \Big|_0^{\cos^{-1}\frac{x}{R}}}{- \cos\theta \Big|_0^{\pi}}$$

$$= \frac{1}{2} (1 - \frac{x}{R})$$

Thus the fraction of all betas escaping from surface, i.e., the average fraction averaged over x = 0 to x = x_{max} = R, is

$$\overline{F} = \frac{\int_0^R F dx}{\int_0^R dx} = \frac{1}{2R} \int_0^R (1 - \frac{x}{R}) dx$$

$$= \frac{1}{2R} (x - \frac{x^2}{2R}) \Big|_0^R$$

$$= \frac{1}{2R} (R - \frac{1}{2}R) = \frac{1}{4}$$

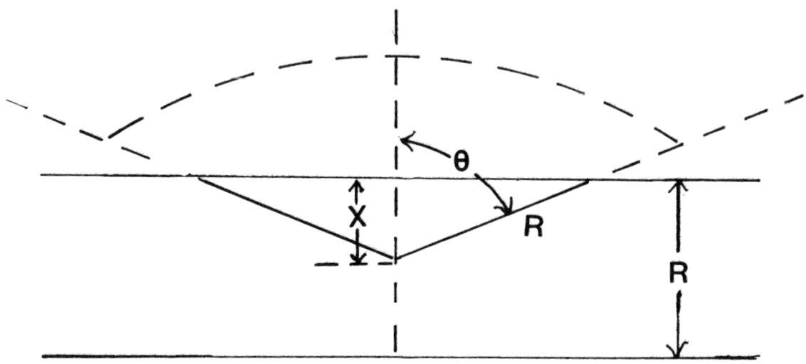

FIGURE B.1 Cone of beta emission from depth x.

APPENDIX C: VARIOUS CONSTANTS

Proton Rest Mass	1.007277 amu
Neutron Rest Mass	1.008665 amu
Electron Rest Mass	0.000548597 amu
Avogadro's Number	6.02252×10^{23} molecules/gm mole (atoms/gm atom)
Rydberg Constant (infinite)	109,737.309 cm^{-1}
Boltzman's Constant	1.38054×10^{-16} erg K^{-1} 8.617065×10^{-5} eV K^{-1}
Speed of Light	2.997925×10^{10} cm/sec
Planck's Constant	6.6252×10^{-27} erg sec 4.1355×10^{-15} eV sec
Electron Charge	1.602×10^{-19} Coul

APPENDIX D: USEFUL CONVERSION FACTORS

1 amu	1.660438×10^{-24} gm
	931.482 MeV
1 Curie	3.7×10^{10} disintegrations/sec
1 esu	1 $gm^{1/2}$ $cm^{3/2}$/sec
1 Coulomb	2.998×10^{9} esu
1 erg	1 gm cm^2/sec^2
	0.6242×10^{6} MeV
1 eV	1.60210×10^{-12} erg
	1.517×10^{-22} Btu
	4.44×10^{-26} kW hr
1 fission	200 MeV (total)
	8.9×10^{-18} kW hr (total)
	180 MeV (in fuel)
1 Joule	10^{7} ergs
1 hp	2545 Btu/hr
	0.7457 kW
	550 ft lb/sec
1 unit electronic charge	4.80×10^{-10} esu
1 Watt	1 Joule/sec
	3.4 Btu/hr

INDEX